OrCAD® PSpice® for Windows

Volume III: Digital and Data Communications

Third Edition

Roy W. Goody
Mission College

Upper Saddle River, New Jersey
Columbus, Ohio

Vice President and Publisher: Dave Garza
Editor in Chief: Stephen Helba
Acquisitions Editor: Scott J. Sambucci
Production Editor: Rex Davidson
Design Coordinator: Robin G. Chukes
Cover Designer: Becky Kulka
Cover Art: FPG
Production Manager: Pat Tonneman
Marketing Manager: Ben Leonard

The book was printed and bound by Victor Graphics, Inc.
The cover was printed by Victor Graphics, Inc.

OrCAD® and PSpice® are registered trademarks of Cadence Design Systems.

Portions of this text previously published as *MicroSim® PSpice® for Windows, Volume II: Operational Amplifiers and Digital Circuits*, copyright 1998 and 1996 by Prentice-Hall, Inc.

Prentice
Hall

10 9 8 7 6 5 4 3 2 1

ISBN   0-13-031183-9

Contents

Preface

OrCAD®PSpice® for Windows, Volume III is the third manual in a three-volume series on PSpice circuit simulation. It covers Digital and Data Communications. (Volume I covers DC/AC Circuits and Volume II covers Devices, Circuits, and Operational Amplifiers.)

If you are already familiar with Volume I or II, you will find few surprises in Volume III. If not, here is a brief overview: The text is based on the latest free evaluation version 9 of the most popular simulation software on the market today: OrCAD's *PSpice*. It is introductory in nature and is appropriate for those with little or no experience in circuit simulation. The level of difficulty is tailored to the technology student, but it offers enough "gentle" material for the technician and enough challenging material for the engineer. It covers both PSpice techniques and electronic theory and applications, and the choice and sequence of material closely follows that of a conventional text on digital. Since most activities can be done by either PSpice simulation or hands-on construction, it is designed to replace a conventional laboratory manual.

Why PSpice for Windows?

As a devoted educator, the chances are you strongly believe that your students must have a circuit simulation experience before they apply their knowledge and skills on the job. Would it not be reasonable to seek out the most popular circuit simulation software on the market today? Would it not also be highly beneficial to use the same software package that is used by engineers and technicians on the job—restricted only by circuit complexity? And would it not be sensible to reduce your costs to zero, and to distribute the software without regard for licenses or copyright restrictions?

Volume III

We assume that the majority of students reading this preface have at least a passing familiarity with Volume I or II. The majority of the most fundamental PSpice techniques and processes are covered in Volume I. Since there is insufficient room to repeat all the introductory material here, short review inserts are presented where appropriate and special Simulation Notes are reprinted in Appendix A. Therefore, if this text is being used in a classroom situation, we strongly recommend that several copies of Volume I be available for reference and review.

Volume III is divided into six parts: Gates and Flip-Flops; Violations and Hazards; Counters, Shift Registers, Coders, and Timers; Converters and RAM; Data Communications; and Modular Design and Applications. This is generally the same order and mix of subjects found in a typical digital course. The modular techniques of Part 6 can, for the most part, be applied to any of the previous chapters at any time.

OrCAD's Total Solution

For designing electronic circuits, OrCAD offers a total solution package, including schematic entry, FPGA synthesis, digital, analog, mixed-signal simulation, and printed circuit board layout—everything from start to finish. All software components are fully integrated and are designed to follow an engineer's natural design flow.

This text is based exclusively on just one part of the complete package: PSpice A/D. Fortunately, PSpice A/D is precisely what we need to support a college-level technology class, for this software component simulates nearly any mix of analog and digital circuits and conveniently displays the results in graphical form. It is incredibly powerful, easy to learn, and simple to use. Quite simply, OrCAD's PSpice A/D is one of the best learning tools available.

OrCAD Lite

Fortunately, for those of us in education, OrCAD Corporation has made PSpice evaluation software available at no cost. All the activities in this book are based on OrCAD Lite version 9.2. Its only major limitation is the number of symbols and components that can be placed on the schematic. Fortunately, we can adjust easily to these limitations, and for the most part they will be completely invisible.

Newer versions of the OrCAD software are constantly being released and the chances are good that they will work with this manual. In general, you should use the latest version that is available; if any adjustments are necessary, they should be minor. *However, to be perfectly safe, you may wish to stay with version 9.2 until the next edition of PSpice for Windows is released.*

Suggestion

Although circuit simulation is the major design and development tool of the future, we recommend that the reader also receive hands-on experience by prototyping actual circuits and troubleshooting with conventional instruments.

One computer-saving approach is to divide a class into two or more groups and switch between PSpice and hands-on techniques. It is especially instructive to perform the same activity using both PSpice and hands-on techniques, and to compare the two approaches. *In this regard, most of the experimental activities outlined in this text can be performed using either PSpice or hands-on techniques.*

Further Study

If you order the complete set of manuals that comes with PSpice A/D, you would be confronted with more than one thousand pages of data, instructions, and reference material.

Clearly, all the information contained within those pages cannot be placed into this introductory text series. Instead, we have included only the most vital and commonly used features of PSpice. For a comprehensive description of all the features of PSpice, refer to the complete set of manuals from OrCAD.

Product manuals and many other useful items and features, including technical data, articles, techtips, and university support, can be obtained from OrCAD's website (www.orcad.com).

Mouse Conventions

Throughout this text, we will adopt the following mouse convention:
- **CLICKL** or **BOLD PRINT** (*click left once*) to select an item.
- **DCLICKL** (*double click left*) to perform an action.
- **CLICKR** (*click right once*) to open a menu.
- **CLICKLH** (*click left, hold down, and move mouse*) to drag a
                selected item. Release left button when placed.

Acknowledgments

I wish to express my sincere gratitude to production editor Rex Davidson and acquisitions editor Scott Sambucci of Prentice Hall. Under their careful guidance, the project steadily moved forward and was released on time.

Of course, OrCAD Corporation deserves special credit for making the OrCAD Lite evaluation version available at no cost. Their foresight makes it possible for colleges and universities to teach circuit simulation at the professional level without breaking the ever-shrinking budget.

Thank you for adopting *OrCAD PSpice for Windows*; May you have good luck and success.

Roy W. Goody

How to obtain your free OrCAD Lite software

During the writing of this text the author relied on PSpice Beta version 9.2. Unfortunately, the free Lite version 9.2 CD (which the circuits of this text are designed for) was not available when this text went to the presses—but should be by the time it reaches your hands.

   To obtain your free copy of PSpice Lite do any of the following:

- Go to OrCAD's website (www.orcad.com) and order the software as a CD to be delivered by mail.

- Download the software directly from OrCAD's website.

- Phone OrCAD sales at 1-888-671-9500.

- Instructors and professors can obtain a free copy of demo 9.2 by simply ordering the *PSpice for Windows Instructor's Guide* from Prentice Hall (www.prenticehall.com) or by calling your local representative.

Part 1

Gates and Flip-Flops

Logic gates are the cells of the digital world—the fundamental building blocks of all digital circuits, from simple logic machines to complex microprocessors.

In the three chapters of Part 1 we introduce the basic logic gates, see how they are stimulated, and finally how they are combined to form the prime unit of computer memory (the flip-flop).

CHAPTER

1

TTL
Logic Gates

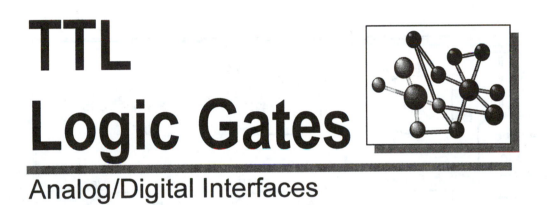

Analog/Digital Interfaces

Objectives

- *To determine the specifications of the TTL logic family*
- *To determine how PSpice interfaces analog and digital components*
- *To construct and test a variety of simple logic circuits*
- *To display combined analog and digital waveforms*

Discussion

The oldest, most widely used logic family is TTL (transistor-transistor-logic).   In the ideal case, a TTL gate is simply a device that performs Boolean algebra operations.   In the real world, however, it has threshold voltages, propagation delays, loading factors, and other vital performance characteristics.   These specifications are the subject of this chapter.

TTL Gates

Figure 1.1 shows that all six basic TTL gates are available to us when using the Lite version of PSpice.

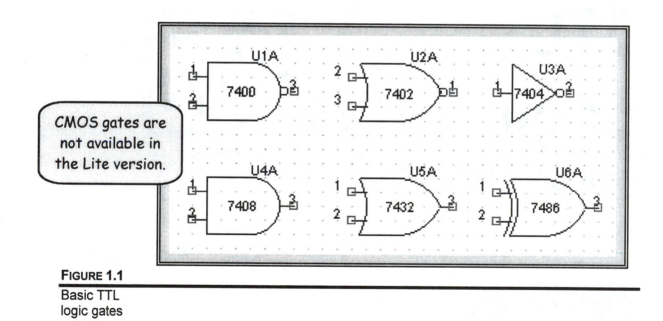

FIGURE 1.1

Basic TTL
logic gates

Analog-to-Digital Interfacing

PSpice allows us to combine analog and digital devices in any way we wish.   In most cases, it is not necessary for the designer to know the behind-the-scenes details concerning such mixed circuits. However, when displaying nodal values, a more detailed knowledge may prove essential.

Analog, Digital, and Interface Nodes

PSpice evaluates three types of nodes: *analog*, *digital*, and *interface*. If all the devices connected to a node are analog, then the node is analog. If all the devices are digital, then the node is digital. If both analog and digital devices drive the node, then it is an interface node. For each type of node, certain *values* are assigned:

- For analog nodes, the values are voltages and currents.

- For digital nodes, the values are states, which are calculated from the input/output model for the device, the logic level of the node (0 or 1), and from the output strengths of the devices driving the node. Each strength is one of 64 ranges of impedance values between the default of 2 and $20k\Omega$. To find the strength of a digital output, PSpice uses the logic level and the values of parameters DRVH (high-level driving resistance) and DRVL (low-level driving resistance) from the device's I/O models.

- For interface nodes, the values are both analog voltages and currents and digital states. This is because PSpice automatically inserts an AtoD or DtoA *interface subcircuit* (complete with power supply) at all interface nodes. These subcircuits handle the translation between analog voltages and currents and digital states.

Figure 1.2 shows a simple mixed analog/digital circuit. Although this is the way the circuit is drawn under Schematics, the interface nodes do not show the invisible circuitry automatically added by PSpice. For that we turn to Figure 1.3, where we have simply inserted the interface blocks.

From Figure 1.3 we note the two interface circuits (AtoD and DtoA) created automatically by PSpice. With the wire segments labeled as shown, there are two analog voltage nodes [*V(Vin)* and *V(Vout)*] and three digital voltage nodes (*Vin$AtoD*, *Vmiddle*, and *Vout$DtoA*). As we will see, Probe (the software component that displays waveforms) handles digital values quite differently from analog values. That is, analog values will resemble oscilloscope waveforms, and digital traces will have the squared-off look of the logic analyzer display.

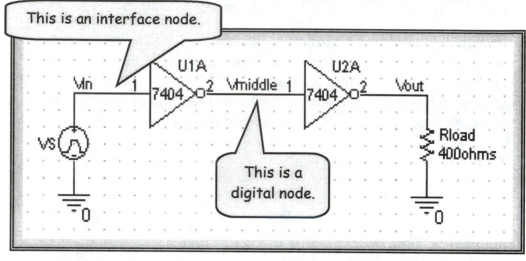

FIGURE 1.2

Mixed component
circuit

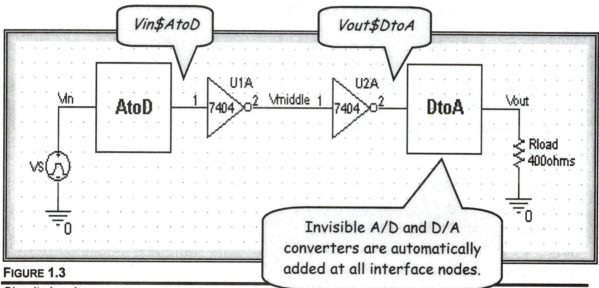

FIGURE 1.3

Circuit showing
invisible nodes and
interface devices

Although the interface circuits do not appear in the schematic, their characteristics are detailed in the output file—as listed in Figure 1.4. (Note that PSpice also automatically creates a single interface power supply and ground for the AtoD and DtoA subcircuits.)

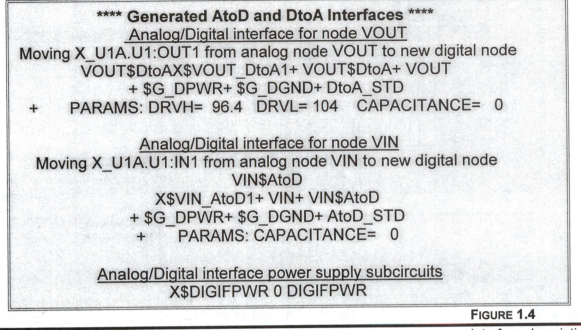

```
**** Generated AtoD and DtoA Interfaces ****
     Analog/Digital interface for node VOUT
Moving X_U1A.U1:OUT1 from analog node VOUT to new digital node
       VOUT$DtoAX$VOUT_DtoA1+ VOUT$DtoA+ VOUT
           + $G_DPWR+ $G_DGND+ DtoA_STD
  +    PARAMS: DRVH= 96.4  DRVL= 104   CAPACITANCE=  0


       Analog/Digital interface for node VIN
Moving X_U1A.U1:IN1 from analog node VIN to new digital node
                    VIN$AtoD
          X$VIN_AtoD1+ VIN+ VIN$AtoD
          + $G_DPWR+ $G_DGND+ AtoD_STD
          +    PARAMS: CAPACITANCE=  0


  Analog/Digital interface power supply subcircuits
              X$DIGIFPWR 0 DIGIFPWR
```

FIGURE 1.4

Interface description
in the output file

In this chapter we will use several analog/digital circuits to study the basic characteristics and specifications of analog, digital, and interface nodes.   Many of our results will be compared to the specification sheet values of Appendix C.

Simulation Practice

Activity *INVERTER*

Activity *INVERTER* of Figure 1.5 will be used to investigate the properties of TTL circuits and analog/digital interfacing.

1. Create project *logicgates* with schematic *INVERTER*.

2. Draw the test circuit of Figure 1.5.

Create Document

PSpice for Windows

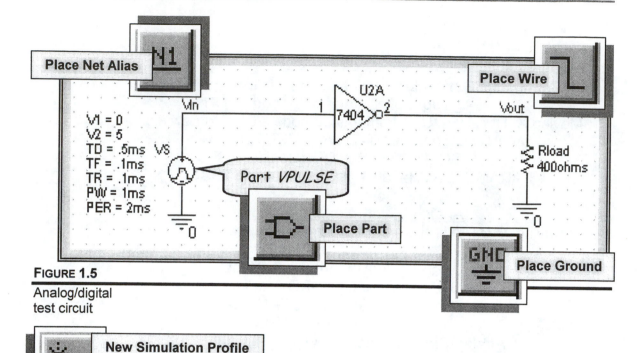

FIGURE 1.5

Analog/digital
test circuit

3. Set the simulation profile to Transient from 0 to 2ms (step ceiling of 2μs). Run a simulation, generate the default Probe graph, and bring up the Add Traces dialog box.

   a. Select various combinations of **Analog**, **Digital**, **Voltages**, **Currents**, **Power**, **Alias Names**, and **Subcircuit Nodes** and note the sets of variables that are available in each case.

   b. For each case below, list only the variable categories that *must* be enabled to display each variable.

   - V(VIN)

   - I(Rload)

   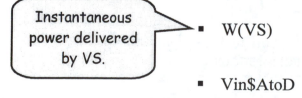

   - W(VS)

   - Vin$AtoD

4. Using *analog* trace names, generate the input/output *analog* voltage and current curves of Figure 1.6.

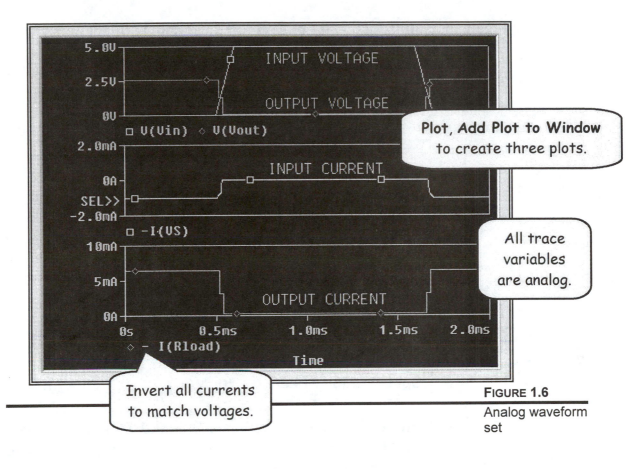

FIGURE 1.6
Analog waveform set

In the following section, we will determine a number of TTL specifications. If time is short, begin with those of greatest interest. (Be sure to use the cursors.)

Toggle Cursor

5. Based on the results of Figure 1.6, fill in the following voltages and currents in the tables provided and compare them to the 7404 specifications given:

- **Output high and low voltages (V_{OH} and V_{OL}):** *minimum and maximum logic 0 and 1 output voltages.* (What is V_{OUT} when V_{IN} is 0V and +5V?)

	V_{OH}	V_{OL}
PSpice		
Specification Sheet	2.4V (min)	0.4V (max)

- **Input high and low voltages (V_{IH} and V_{IL})** — *the minimum and maximum input voltages that are guaranteed to represent the input high and input low states.* (What values of V_{IN} will give a V_{OUT} of .4V and 2.4V?)

	V_{IH}	V_{IL}
PSpice	_____	_____
Specification Sheet	2.0V (min)	0.8V (max)

- **Input high and low currents (I_{IH} and I_{IL}):** *maximum input currents at the high and low input states.* (What is I_{IN} when V_{IN} is 5V and 0V?)

$I_{IN} = -I(VS)$

	I_{IH}	I_{IL}
PSpice	_____	_____
Specification Sheet	40μA (max)	1.6mA (max)

- **Output high and low currents (I_{OH} and I_{OL}):** *typical logic 1 and 0 output currents.* (What is I_{OUT} when V_{OUT} is above 2.4V and below .4V?)

$I_{OUT} = -I(Rload)$

	I_{OH}	I_{OL}
PSpice	_____	_____
Specification Sheet	16mA	400μA

Output Short-Circuit Current (I_{OS})

6. Short the output (reduce *Rload* to .001ohms) and generate the output *analog* current plot of Figure 1.7. Based on the results, fill in the output short circuit current for the output high state (V_{IN} = 0V) in the table below. (When done, return *Rload* to 400Ω.)

	I_{OS}
PSpice	_____
Specification Sheet	25mA

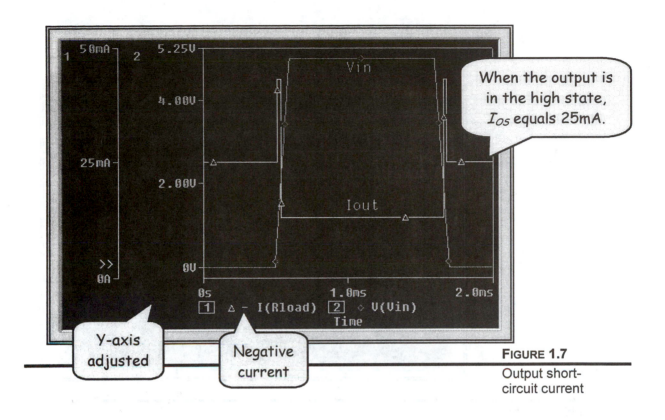

FIGURE **1.7**
Output short-circuit current

Propagation Delay

7. Modify the test circuit and the simulation profile to match that of Figures 1.8 and 1.9.

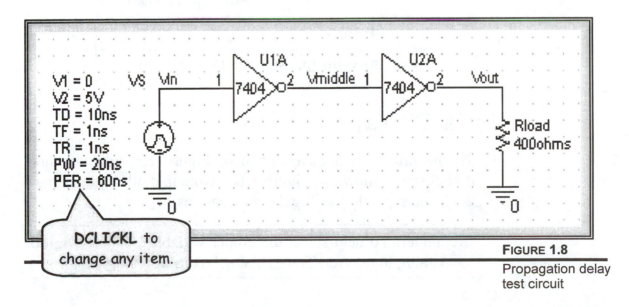

FIGURE **1.8**
Propagation delay test circuit

PSpice for Windows

8. Run PSpice and generate the input/output *analog* waveforms of Figure 1.9.

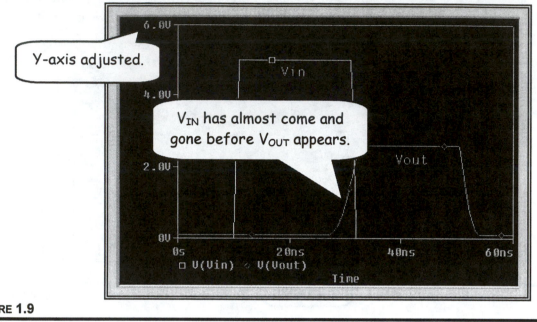

FIGURE 1.9

Propagation delay test results

9. Using any two corresponding points, fill in the value below. (*Reminder*: The signal has passed through two gates.)

	Propagation Delay
PSpice	_____
Specification Sheet	15ns (typical)

Digital Displays

10. To the propagation delay graph of Figure 1.9, add the three digital traces shown in Figure 1.10. Note that all digital signals are listed separately as logic state diagrams.

> Reminder: To help select the proper variables, enable **Digital** or **Analog** in the Add Traces dialog box as necessary.

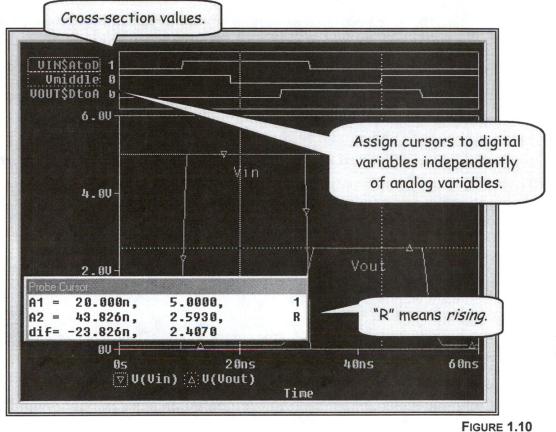

FIGURE 1.10

Adding a purely
digital node voltage

11. Enable the cursor and note the cursor window (as shown in Figure 1.10).

> *Reminder*: Use arrows to move cursor 1, and Shift/arrows to move cursor 2.

12. To gain experience in the use of cursors on mixed analog/digital waveforms, refer to Figure 1.10 and answer as many of the following as you can.

   a. Do the vertical axes of both cursors extend through both the analog and digital sections of the display?

      Yes   No

b. Locate the cross-section values right after the digital trace variables near the top of the screen. Move both cursors and note the values. Are the cross-section values based only on cursor 1?

Yes   No

Does state R seem to indicate rising and state F falling?

Yes   No

c. Locate the cursor window. Does it have an analog section and a digital section?

Yes   No

d. Using **CLICKL** (for cursor 1) and **CLICKR** (for cursor 2), select various analog and digital trace variables. Can the analog and digital trace variables be independently selected?

Yes   No

e. Set up the cursor system as shown in Figure 1.10. Within the cursor window, do the 1 and R correctly reflect the states of the digital waveforms?

Yes   No

f. Associate cursor 1 with analog trace $V(Vin)$ and digital trace $Vin\$AtoD$. Starting at 10ns, move the cursor to the right and note both the digital and analog readings. Did the R (rising) state occur approximately between analog values 1.5V and 2V?

Yes   No

Does this agree with the input HIGH and LOW voltages (V_{IH} and V_{IL}) of step 5?

Yes   No

13. Select various digital and analog trace variables, move the cursors, and note the displays until you become familiar and comfortable with the overall cursor system.

Digital Zoom Techniques

14. Referring to *Simulation Note 1.1,* place zoom bars as shown in Figure 1.11 and zoom in on the selected area to generate the waveform of Figure 1.12.

    a. Are both the analog and digital sections expanded in unison?

        Yes   No

    b. Reactivate the cursor system.   Does the parallelogram within the fall time correspond to the F region?

        Yes   No

These parallelograms are also known as *ambiguity regions* and will be a major topic of Chapter 4.

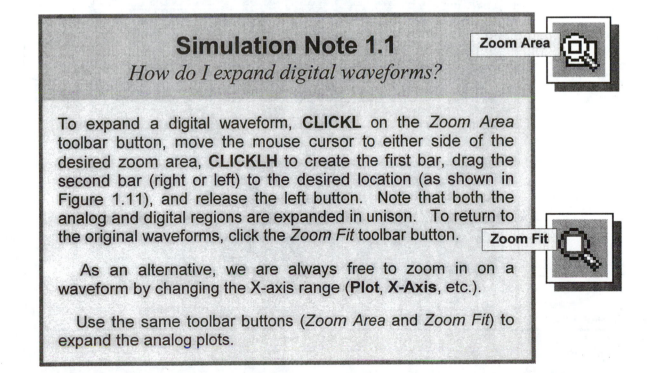

Simulation Note 1.1
How do I expand digital waveforms?

Zoom Area

To expand a digital waveform, **CLICKL** on the *Zoom Area* toolbar button, move the mouse cursor to either side of the desired zoom area, **CLICKLH** to create the first bar, drag the second bar (right or left) to the desired location (as shown in Figure 1.11), and release the left button. Note that both the analog and digital regions are expanded in unison. To return to the original waveforms, click the *Zoom Fit* toolbar button.

Zoom Fit

    As an alternative, we are always free to zoom in on a waveform by changing the X-axis range (**Plot**, **X-Axis**, etc.).

    Use the same toolbar buttons (*Zoom Area* and *Zoom Fit*) to expand the analog plots.

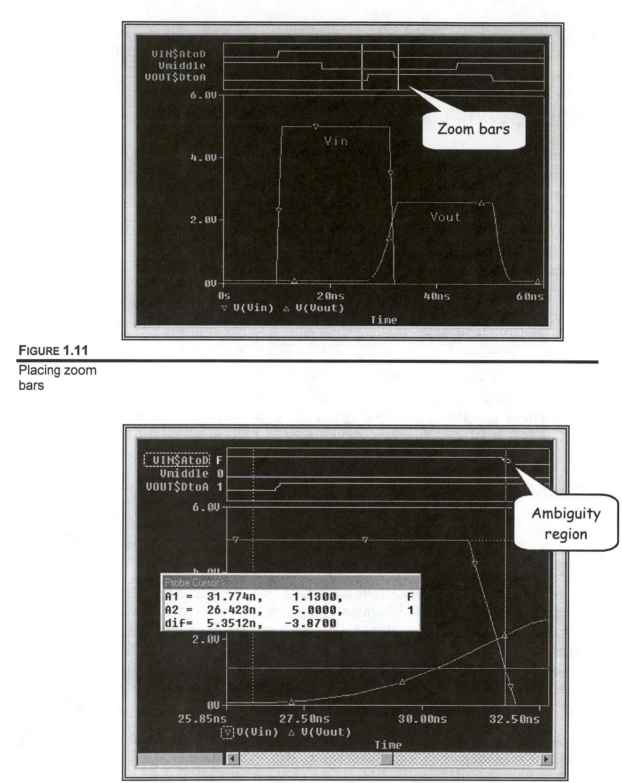

FIGURE 1.11

Placing zoom
bars

FIGURE 1.12

Expansion to
zoom bars

PSpice for Windows

15. Examine *Simulation Note 1.2* and change the size of the digital display from 33% of the total waveform space to 50%. (When finished, return to 33%.)

Simulation Note 1.2
How do I change the size of the digital display?

To change the size of the digital display, **Plot**, **Digital**, **Size**, change the *Percentage of Plot to be Digital* from 33 to 50, **OK**.

To increase the trace name space, increase the default *Length of Digital Trace Names*.

Both of the above can be changed by placing the cursor at the appropriate boundary and dragging.

Markers in a Digital Circuit

16. Place markers on our test circuit as shown in Figure 1.13, and re-run PSpice to generate the waveforms of Figure 1.14.

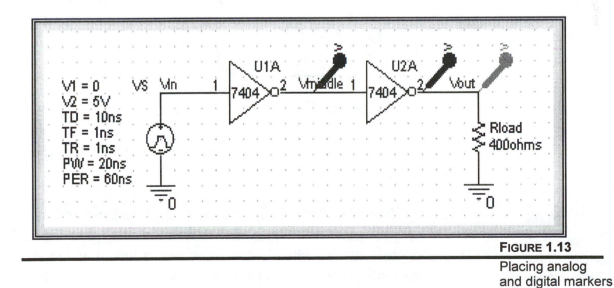

FIGURE 1.13

Placing analog and digital markers

17. Viewing the results (Figure 1.14):

    a. Are digital waveforms generated only when markers are placed at digital nodes?

            Yes   No

    b. Are analog waveforms generated when markers are placed at *either* analog or mixed nodes?

            Yes   No

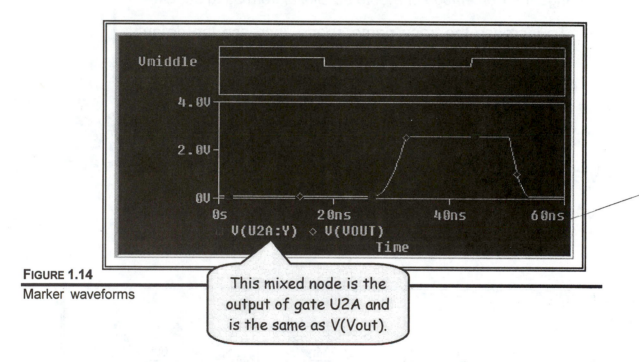

FIGURE 1.14
Marker waveforms

This mixed node is the output of gate U2A and is the same as V(Vout).

Advanced Activities

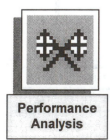

Performance Analysis

18. From step 4, we find that the minimum allowed logic 1 output voltage level is 2.4V.

    Set up *Rload* (of Figure 1.5) as a parametric variable and use the **Max** goal function to generate the graph of Figure 1.15. Based on this graph, what is the minimum allowed value of *Rload* that will guarantee a logic 1 output voltage? Is the value below 400Ω? (See Volume II, Chapter 30, for a complete discussion of performance analysis.)

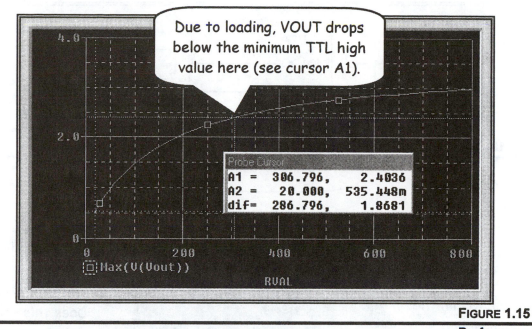

FIGURE 1.15

Performance analysis of maximum *Vout* versus *RL*

19. Returning to Figure 1.5 perform a parametric sweep of temperature (perhaps from −50 to +50, in increments of 25) and display the waveforms. Does temperature influence the analog and digital waveforms in any significant manner?

20. View traces of the interface power supply [V($G_DGND) and V($G_DPWR)]. Are the results as expected?

Exercises

1. Add the necessary source and load, and test the *clock-edge detection* circuit of Figure 1.16.

2. Using analog signals, generate the transfer functions of Figure 1.17 for the 7404 inverter and 7414 Schmitt trigger. (What is *hysteresis*?)

3. Test the circuit of Figure 1.18 with VDC sources and state why it is called a *data selector*.

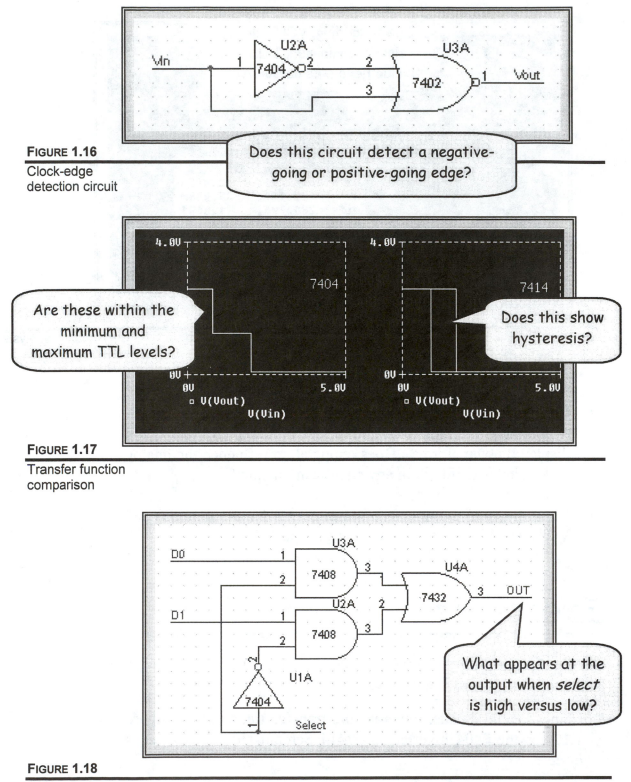

4. Determine the frequency of oscillation of the circuit of Figure 1.19. (*Hint*: The period is approximately 30ns.)

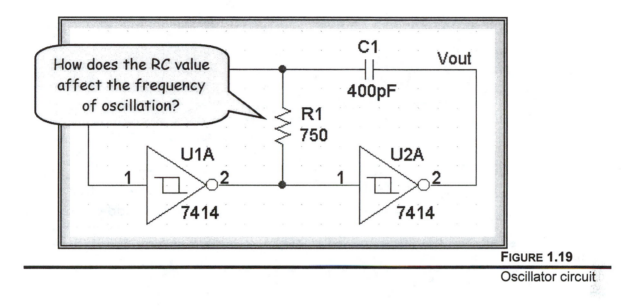

FIGURE 1.19

Oscillator circuit

5. Add the necessary sources and test the full adder of Figure 1.20. (Verify that A = 1 + B = 0 + C$_{IN}$ = 1 equals 0 with a carryOUT of 1.)

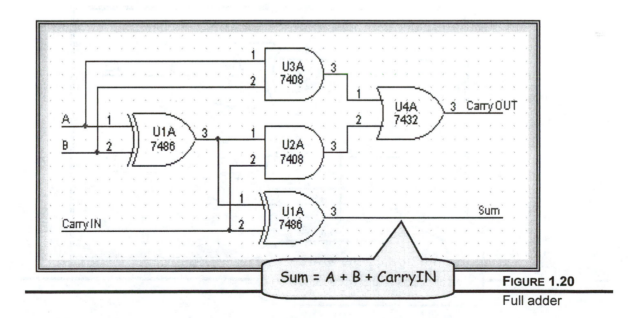

FIGURE 1.20

Full adder

6. Add the necessary sources and test the 4-bit even-parity generator of Figure 1.21.

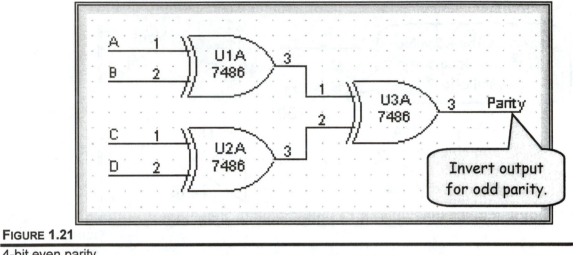

FIGURE 1.21

4-bit even parity generator

7. What mathematical operation does the 2-bit circuit of Figure 1.22 perform? (Could the process be extended to any number of bits?)

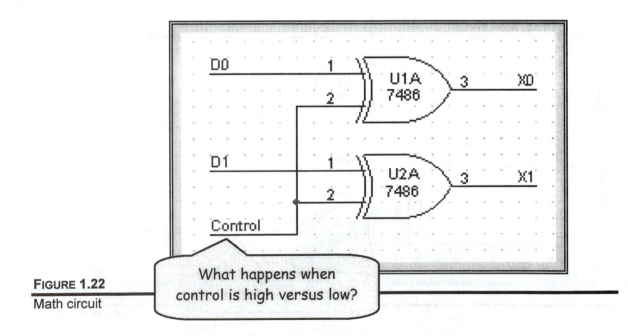

FIGURE 1.22

Math circuit

8. What is the purpose of the mystery circuit of Figure 1.23? (Under what conditions is the output high?)

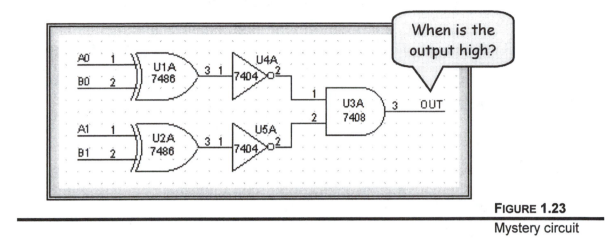

FIGURE 1.23

Mystery circuit

Questions and Problems

1. What are the three possible types of nodes in a mixed analog/digital circuit?

2. Where does PSpice automatically place an AtoD interface circuit? A DtoA interface circuit?

3. Can we measure current at a purely digital node?

4. Regarding the digital gate indicators below, what do the terms "A" and "Y" (to the right of the colon) indicate?

         V(U1A:A)   V(U2A:Y)

5. What is an *ambiguity region*?

6. Explain how the 7414 transfer function of Figure 1.17 shows the effects of *hysteresis*.

CHAPTER

2

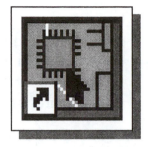

Digital Stimulus

File Stimulus

Objectives

- *To generate a variety of digital waveforms using the digital stimulus generator feature of PSpice*
- *To generate digital waveforms from information stored in a file*

Discussion

In modern computer systems, it is not unusual for the majority of components and interfaces to be purely digital. Quite often, only the initial input and final output are interfaced to analog devices. Furthermore, it is common for digital signals to appear in parallel, that is, in groups of 4, 8, 16, or 32.

For these reasons PSpice provides the routing and stimulation features of Figure 2.1. For routing parallel digital signals, they provided a *bus system*; for generating individual logic levels they gave us the *logic level sources*, and for generating parallel signals, they included the *stimulus generators*.

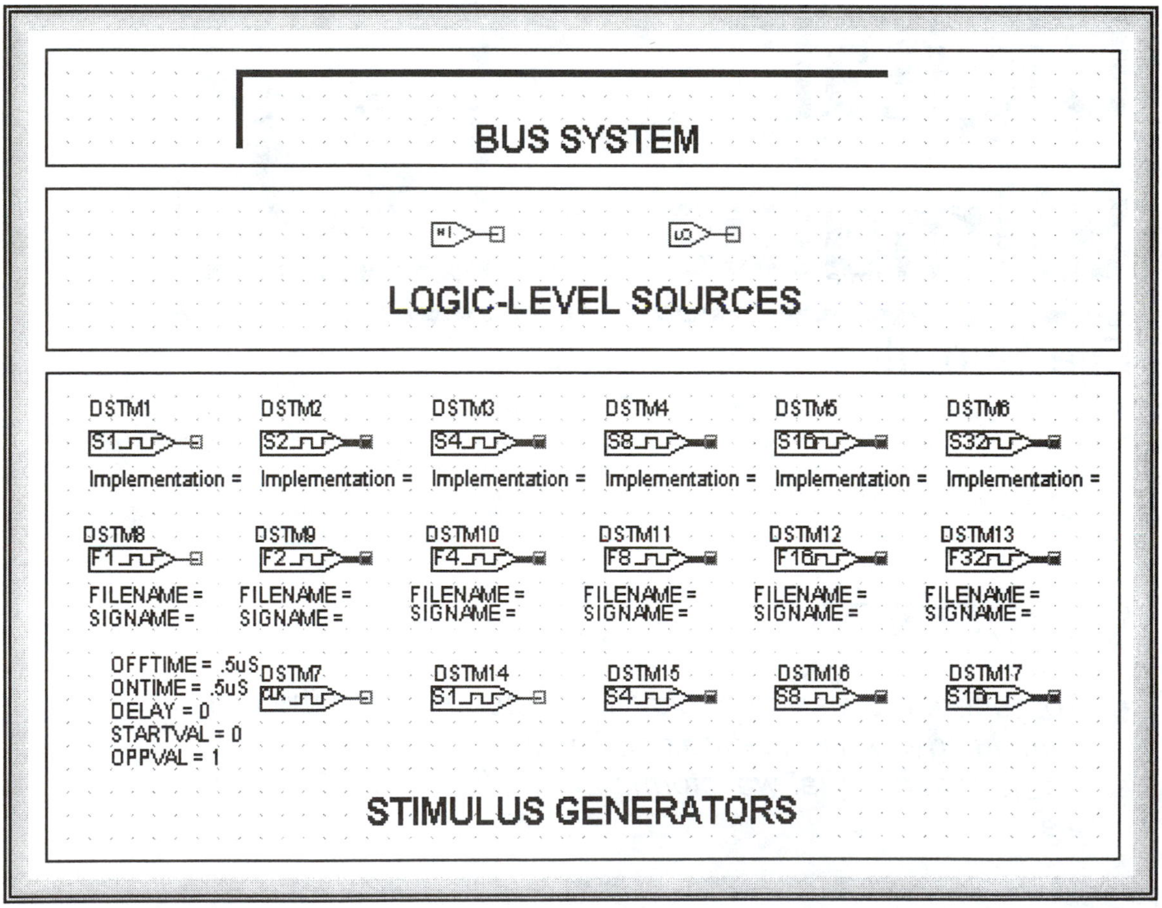

FIGURE 2.1

Logic-level and
digital stimulus devices

System Bus

In a digital system, information is often stored and moved about in *groups* of bits. To simplify the diagramming of such circuits, schematics provides the system bus, a single heavy line drawn on the schematic screen and properly labeled. All bus-system wires are simply connected to this single bus line, and all connections are identified by labels.

The use of a system bus will prove to be especially handy when we perform simulations on computer devices, such as random-access memory (RAM).

Logic-Level Sources

The logic-level sources are straightforward and easy to use. They simply apply a constant logic 1 (+5V for TTL) or logic 0 (0V for TTL) to any single node.

For example, the two-input NAND gate of Figure 2.2 uses the HIGH (+5V) logic level source to permanently enable one input and turn the NAND gate into an inverter. The HI and LO logic level sources will greatly speed up circuit construction and wiring.

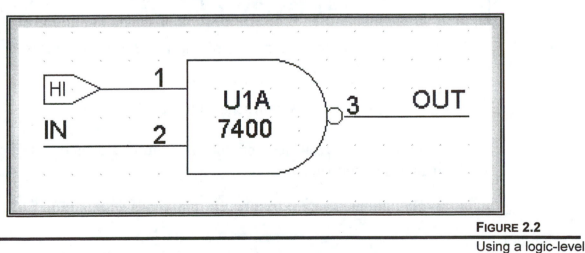

FIGURE 2.2

Using a logic-level source

Digital Stimulus Generators

The digital stimulus generators are divided into four categories:

- The *digStim* devices (top row of the *STIMULUS GENERATORS* section of Figure 2.1) come in six versions (S1, S2, S4, S8, S16, and S32) for bus widths of 1, 2, 4, 8, 16, and 32 lines. They are custom programmed by way of a stimulus editor.   (In the evaluation version, only S1 is available.)

- The *fileStim* devices (second row) are used when the number of stimulus commands is very large. They obtain the stream of digital signals from a file on a hard or floppy disk.

- The *digClock* device (first item, bottom row) generates a continuous clock signal and is directly programmed by setting attributes.

- The *STIM* devices (bottom row, right four) come in four versions (S1, S4, S8, and S16). They are custom programmed by entering command lines. All are available in the evaluation version.

Simulation Practice

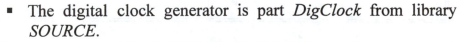

Activity *CLOCK*

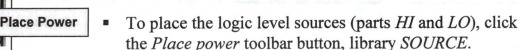

Activity *CLOCK* uses the circuit of Figure 2.3 to generate a continuous clock waveform.

1. Create project *stimulus* with schematic *CLOCK*.

2. Draw the all-digital test circuit of Figure 2.3 and set the attributes as shown.

   - The digital clock generator is part *DigClock* from library *SOURCE*.

   - To place the logic level sources (parts *HI* and *LO*), click the *Place power* toolbar button, library *SOURCE*.

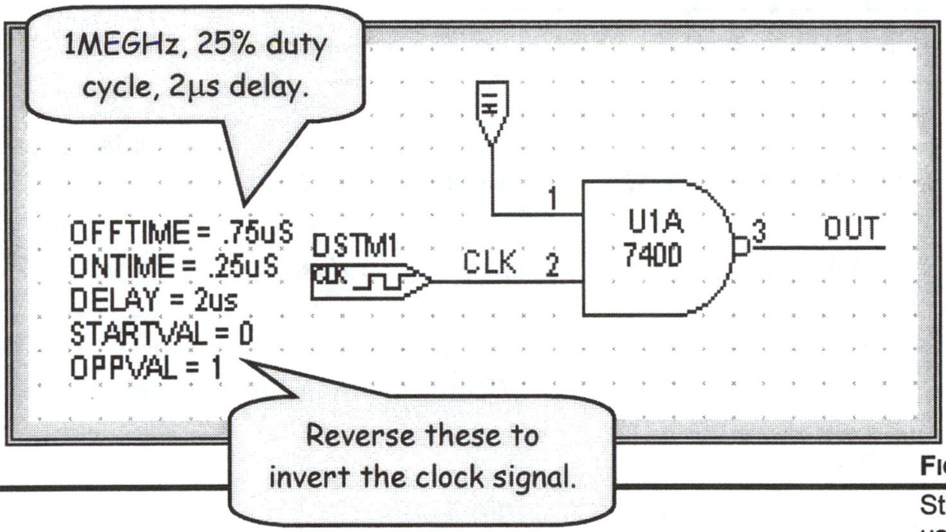

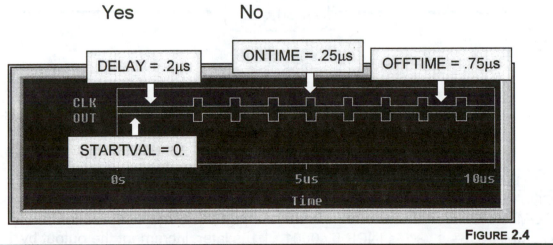

FIGURE 2.3

Stimulus circuit
using digital clock

3. Set the simulation profile to *Transient* from 0 to 10μs, step
    ceiling of .1μs.   Run PSpice and generate the input/output
    waveforms of Figure 2.4.

    a. Were the results as expected?

    > Yes   No

    b. Can the amplitude of the digital signals be measured with the
        cursor?

    > Yes   No

FIGURE 2.4

Clock stimulus
digital waveforms

Activity *STIM4*

Activity *STIM4* uses the circuit of Figure 2.5 to interface a stimulus generator with a small array of gates.

4. Add schematic *STIM4* to project *stimulus*.

5. Using device *STIM4* (from library *SOURCE*), draw the test circuit of Figure 2.5. (Refer to *Simulation Note 2.1* when setting up the bus system.)

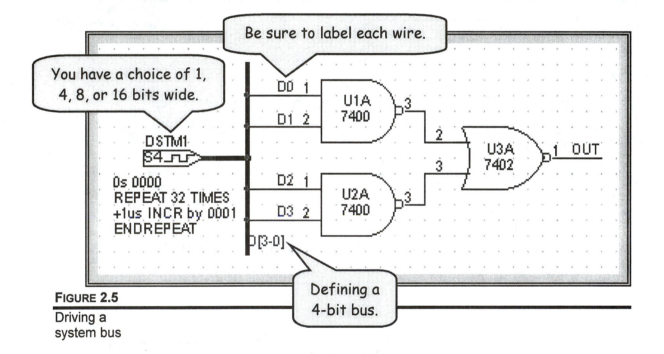

FIGURE 2.5

Driving a
system bus

6. To program *DSTM1*, **DCLICKL** on the symbol, and fill in commands 1 through 4 as shown below. (Display only the value portion of each.)

ITEM	VALUE	DESCRIPTION
COMMAND1=	0s 0000	At 0s, binary D3-D0 = 0000
COMMAND2=	REPEAT 32 TIMES	Repeat 32 times
COMMAND3=	+1μs INCR by 0001	1μs later, increment the output by 1
COMMAND4=	ENDREPEAT	End of repeat loop

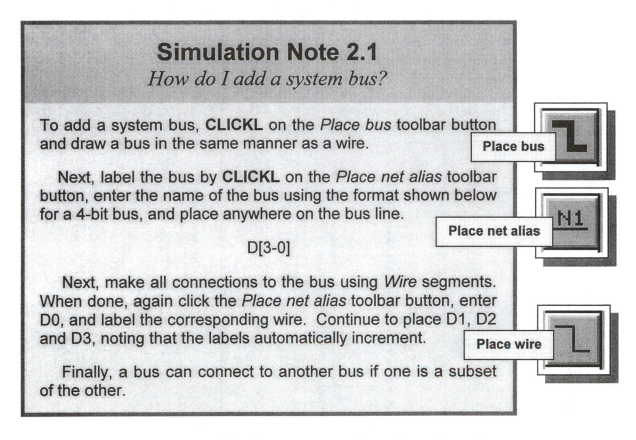

Simulation Note 2.1
How do I add a system bus?

To add a system bus, **CLICKL** on the *Place bus* toolbar button and draw a bus in the same manner as a wire.

Next, label the bus by **CLICKL** on the *Place net alias* toolbar button, enter the name of the bus using the format shown below for a 4-bit bus, and place anywhere on the bus line.

D[3-0]

Next, make all connections to the bus using *Wire* segments. When done, again click the *Place net alias* toolbar button, enter D0, and label the corresponding wire. Continue to place D1, D2 and D3, noting that the labels automatically increment.

Finally, a bus can connect to another bus if one is a subset of the other.

7. Set the simulation profile for *Transient* from 0 to 40μs (step ceiling .4μs), and generate the input/output waveforms of Figure 2.6. Were the results as expected?

Yes   No

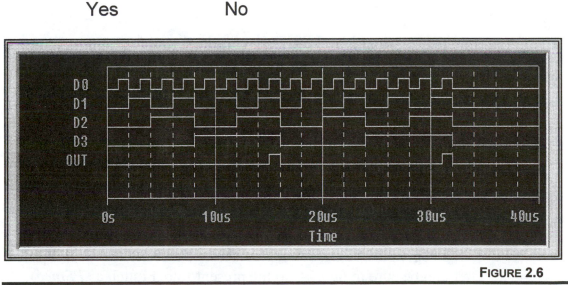

FIGURE 2.6
STIM4 digital waveforms

PSpice for Windows

Activity *FILESTIM4*

Activity *FILESTIM4* uses the circuit of Figure 2.7 to generate a 4-bit digital signal stream from a file. This is especially useful for long non-repetitive waveforms.

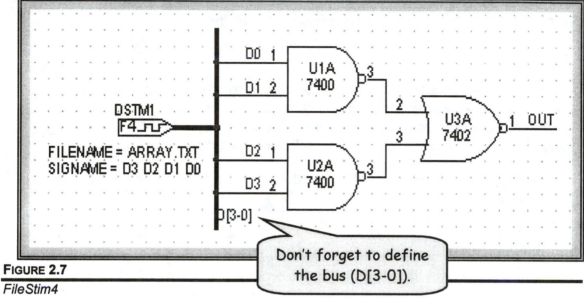

Don't forget to define the bus (D[3-0]).

FIGURE 2.7

FileStim4
stimulus circuit

8. Add schematic *FILESTIM4* to project *stimulus*.

9. Using part *FileStim4* (from library *SOURCE*), draw the test circuit of Figure 2.7, and set the *FILENAME* and *SIGNAME* attributes for *DSTM1* as shown.

10. Using a text editor, open a file (*ARRAY.TXT*) in the current directory, enter the data of Figure 2.8, and close the file.

To make use of Window's text editor (*NotePad*), **Start, Programs, Accessories, NotePad**. Enter the text of Figure 2.8 and be sure to *Save As* to the present folder (such as *PSpice*).

PSpice for Windows

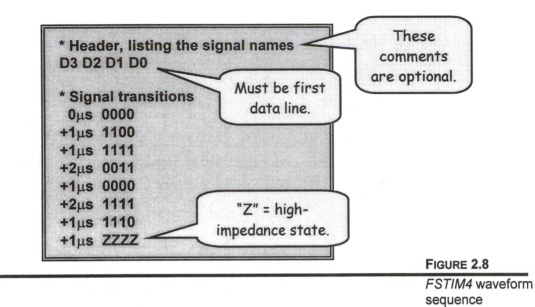

FIGURE 2.8

FSTIM4 waveform sequence

11. Set the simulation profile to *Transient* from 0 to 10μs (step ceiling .1μs), run PSpice, and display the waveform set of Figure 2.9.

   a. Do the input signal data (*D3*, *D2*, *D1*, *D0*) specified by the file (*ARRAY.TXT*) match the corresponding waveforms?

      Yes   No

   b. Does the output signal (*OUT*) match your expectations?

      Yes   No

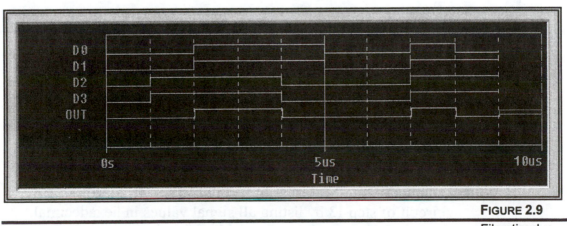

FIGURE 2.9

File stimulus waveforms

Advanced Activities

12. Using the test schematic of Figure 2.10, enter the data below into a file name of your choice, and generate serial waveforms *CLK* and *OUT*. Identify each of the state values (0, 1, R, F, X, and Z) for both *CLK* and *OUT*. (Identify the one time period in which *CLK* and *OUT* differ, and explain the reason for the difference.)

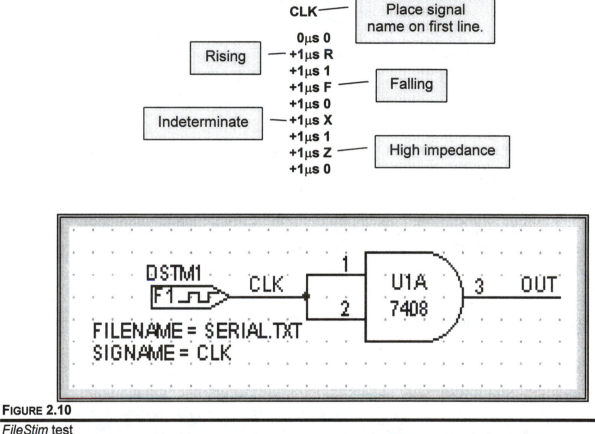

FIGURE 2.10

FileStim test circuit

13. Using either *STIM32* or *FSTIM32*, set up and test a 32-bit stimulus circuit.

14. By modifying the format from 1111 to 4, retest the 32-bit circuit of step 13 by listing all signal values in hexadecimal.

Exercises

1. Substitute 7420 4-bit NAND gates for the 7400s (Figure 2.7), and drive the system with an 8-bit stimulus generator.

2. The 7451 IC of Figure 2.11 consists of two AND and one OR gate internally connected. Using a bus and digital stimulus device of your choice, determine how the gates are arranged inside the IC block.

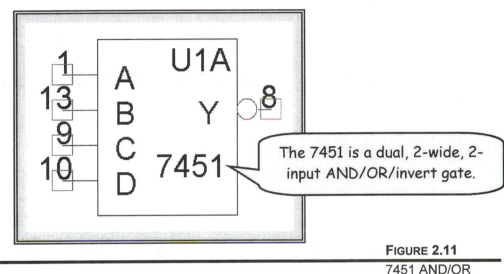

FIGURE 2.11

7451 AND/OR digital IC

Questions and Problems

1. How does the use of a bus simplify digital schematics?

2. What is the digital stimulus symbol for each of the following?

    a. Unknown

    b. High Z

3. What is the purpose of the *FORMAT* parameter? (*Hint*: See step 14.)

4. Under what circumstances would the file stimulation device (*FSTM*) be helpful?

5. Can we place multiple digital stimulus devices on a single circuit?

6. Can we determine the amplitude of purely digital signals?

CHAPTER

3

Latches and Flip-Flops

Edge- Versus Level-Triggered

Objectives

- *To analyze the design of discrete flip-flops*
- *To compare the level-active and edge-triggered D-latch*
- *To demonstrate the characteristics of the JK flip-flop*

Discussion

The flip-flop is a basic unit of data storage (memory). As shown by the circuit of Figure 3.1, a single data bit is stored by positive feedback latching action.

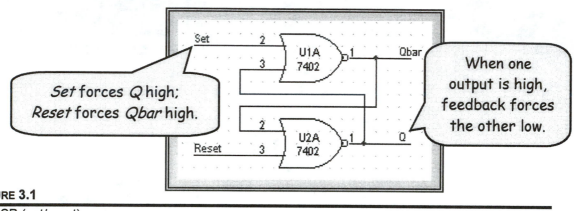

FIGURE 3.1

The SR (set/reset) flip-flop

Synchronous Versus Asynchronous

The set (S) and reset (R) inputs to the SR latch of Figure 3.1 are *asynchronous* (unclocked) because they can occur at any time. By *gating* the *set/reset* inputs, as shown by Figure 3.2, we create a *synchronous* (clocked) flip-flop.

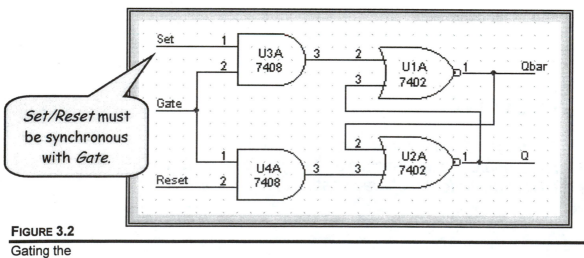

FIGURE 3.2

Gating the input

PSpice for Windows

D-latch

The most basic type of integrated circuit (IC) flip-flop is the D-latch. As shown in Figure 3.3, it comes in two major forms: *edge-triggered* (7474) and *level-triggered* (7475). Table 3.1 compares the synchronous inputs of the 7474 and 7475.

- The synchronous input to the 7474 (similar to the 7475) is edge-triggered rather than level-triggered. That is, with the 7474, data on the D input is transferred to the Q output on the *low-to-high transition* of the clock pulse.

  The active low *preset* (PRE) and *clear* (CLR) signals are asynchronous because they can be activated at any time. They override the clock input and directly set the Q output to 1 and 0.

- The 7475 level-triggered version offers two flip-flops in a single package. Because the outputs (1Q and 2Q) follow the data inputs (1D and 2D) only when the clock (C) is high, the inputs are synchronous.

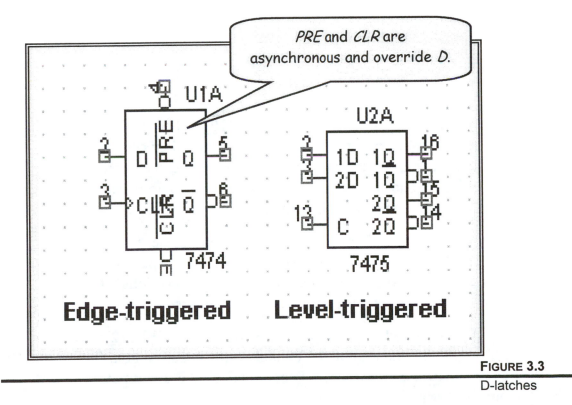

FIGURE 3.3

D-latches

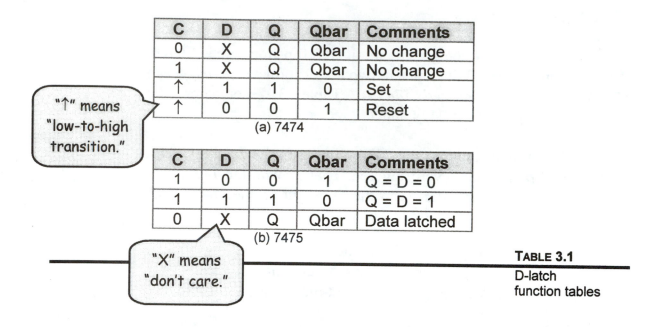

"↑" means "low-to-high transition."

C	D	Q	Qbar	Comments
0	X	Q	Qbar	No change
1	X	Q	Qbar	No change
↑	1	1	0	Set
↑	0	0	1	Reset

(a) 7474

C	D	Q	Qbar	Comments
1	0	0	1	Q = D = 0
1	1	1	0	Q = D = 1
0	X	Q	Qbar	Data latched

(b) 7475

"X" means "don't care."

TABLE 3.1

D-latch function tables

JK Flip-Flop

The JK flip-flop is the most versatile of the three basic types (SR, D, and JK). First and foremost, it offers a new mode of operation: a *toggle* mode, in which the outputs *change* state.

As shown by the examples of Figure 3.4, the JK flip-flop has three major versions:

- Positive pulse-triggered (7473, 7476, and 74107): These older flip-flops are known as *master/slave* devices because they consist of two latches: a master latch to receive data while the input clock goes HIGH, and a slave latch to receive and output data from the master when the clock goes LOW.

- Positive edge-triggered (74109, 74LS109): With these newer devices, the flip-flop latches and outputs JK data upon the positive-going LOW-to-HIGH clock transition. Transitions of the JK inputs before or after the active clock are ignored.

- Negative edge-triggered (74LS73, 74LS76, 74LS107, 74LS112): The flip-flop latches and outputs JK data upon the negative-going HIGH-to-LOW clock transition. Transitions of the JK inputs before or after the active clock are ignored.

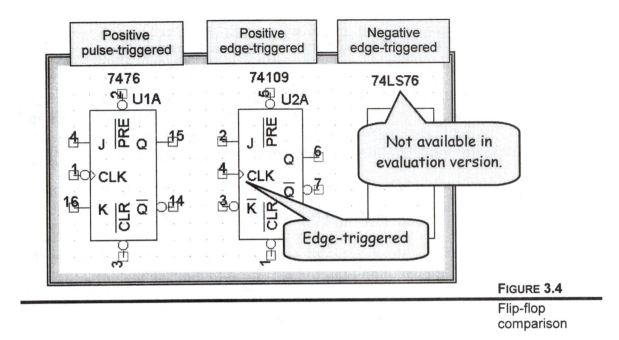

FIGURE 3.4

Flip-flop
comparison

Technically speaking, the small, edge-triggered triangle should not appear on the pulse-triggered 7476.

7476 Master/Slave JK Flip-Flop

As shown by Figure 3.4, the 7476 offers both asynchronous (PRE and CLR) and synchronous (JK) inputs. In Table 3.2, synchronous input (JK) data is loaded into the master upon the positive-going clock transition and transferred to the slave on the negative-going clock transition.

CLK	J	K	Action
⎍	0	0	none
⎍	1	0	set (Q high)
⎍	0	1	reset (Q low)
⎍	1	1	toggle (Q changes)

TABLE 3.2

7476
function table

Simulation Practice

> *Note*: If you experience difficulty with any digital IC (such as receiving only unknowns on the outputs), set the initial state to zero by clicking the Edit Simulation Settings dialog box, **Options**, **Gate-level Simulation**, *Initialize all flip-flops to 0*, **OK**.

Activity *DLATCHL*

Activity *DLATCHL* will investigate the properties of the 7475 level-triggered D-latch of Figure 3.5.

1. Create project *flipflop* with schematic *DLATCHL*.

2. Draw the 7475 D-latch circuit of Figure 3.5.

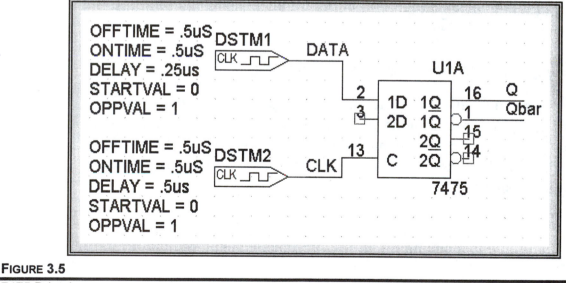

FIGURE 3.5

7475 D-latch
circuit

3. Set the simulation profile for *Transient* from 0 to 4μs, with a step ceiling of 4ns.

4.   When programmed as shown in Figure 3.5, the waveform set of Figure 3.6 shows the input signals to the D-latch.   Below the signals, predict (draw) the Q and Qbar output waveforms in the space provided. (Remember, the 7475 is level-triggered.)

FIGURE 3.6

7475 D-latch
waveforms

5.   Generate the Q digital output waveform using PSpice. Were your predictions correct?   (Make any necessary corrections to your predicted waveform.)

Yes   No

Activity *DLATCHE*

Activity *DLATCHE* will investigate the properties of the 7474 edge-triggered D-latch of Figure 3.7.

6.   Add schematic *DLATCHE* to project *flipflop*.

7.   Draw the circuit of Figure 3.7 and set all attributes as shown.

8.   As with the 7475, sketch your predicted Q and Qbar output waveforms on the graph of Figure 3.8.

PSpice for Windows

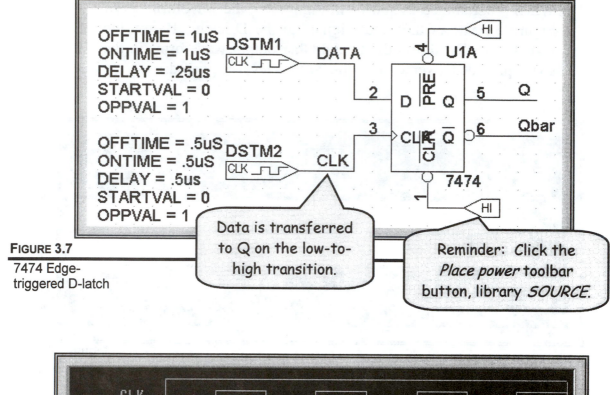

FIGURE 3.7

7474 Edge-
triggered D-latch

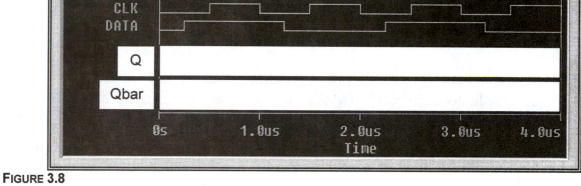

FIGURE 3.8

7474 D-latch
waveforms

9.   Set the simulation profile to *Transient* from 0 to 4μs with a step
     ceiling of 4ns.

10. Generate the Q output waveform using PSpice. Were your
     predictions correct? (Make any necessary corrections to your
     predicted waveform.)

           Yes   No

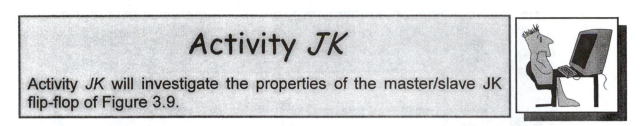

Activity JK

Activity *JK* will investigate the properties of the master/slave JK flip-flop of Figure 3.9.

11. Draw the test circuit of Figure 3.9 and set the attributes as shown.

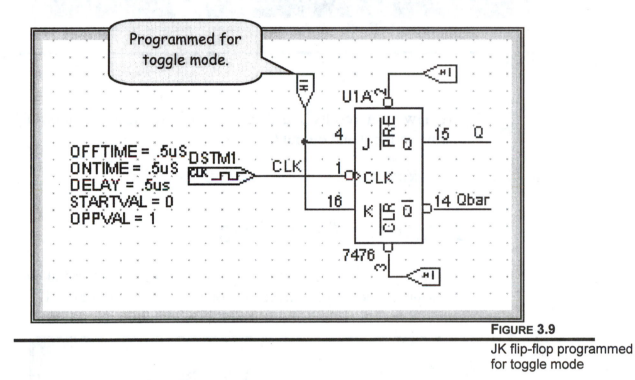

FIGURE 3.9

JK flip-flop programmed for toggle mode

12. As before, sketch your predicted Q and Qbar output waveforms on the graph of Figure 3.10.

13. Set the simulation profile for *Transient* from 0 to 8µs, step ceiling of 8ns.

14. Generate the output waveform using PSpice. Were your predictions correct? (Make any necessary corrections to your predicted waveforms.)

     Yes   No

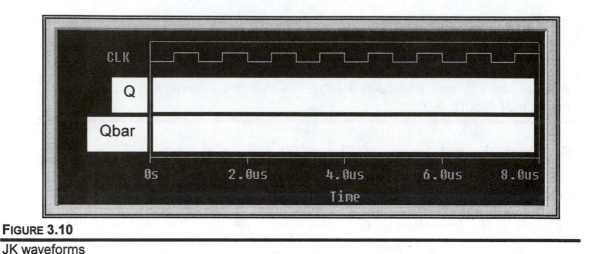

FIGURE 3.10

JK waveforms

15. Review your results and determine the frequency relationship between input and output?

$$f(in) = \underline{\hspace{3cm}} \times f(out)$$

Advanced Activities

16. Modify your 7476 circuit as shown in Figure 3.11, and sketch your predicted output waveforms on the graph of Figure 3.12.

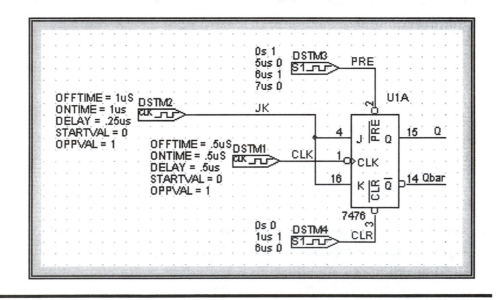

FIGURE 3.11

JK flip-flop
test circuit

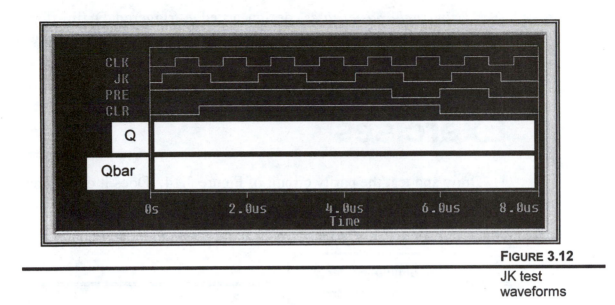

FIGURE 3.12

JK test
waveforms

17. Sketch the Q output for any or all of the brain teaser circuits of Figure 3.13. Use PSpice to test your predictions.

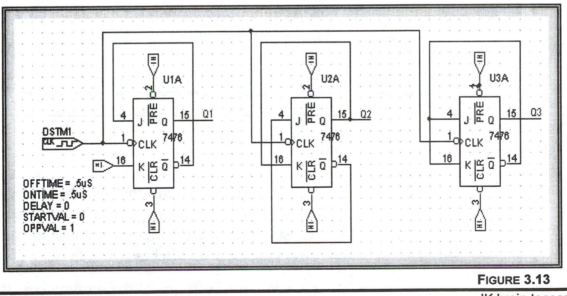

FIGURE 3.13

JK brain teaser
circuits

18. *Using only basic gates*, design an edge-triggered D flip-flop. (*Hint*: Make use of the clock-edge detection circuit of Chapter 1, Figure 1.16.)

Exercises

1. Draw and test the parity circuit of Figure 3.14. Does it determine even or odd parity? (As serial data enters the Data input [*D1*] at the clock rate [*D0*], the circuit keeps a running total of parity.)

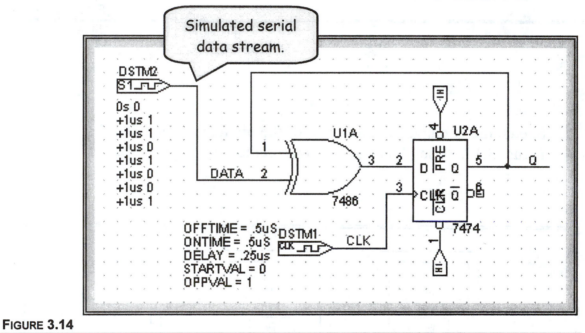

FIGURE 3.14

Parity test circuit

2. Design, draw, and test a "Jeopardy" circuit that has three input switches and three output "lights." During play, whichever switch is activated first causes the corresponding output to go active—and deactivates the other two switches.

3. By examining the structure and waveforms of the mystery circuit of Figure 3.15, determine its function and purpose. (*Hint*: Its initials are "SR," and we don't mean "set /reset.")

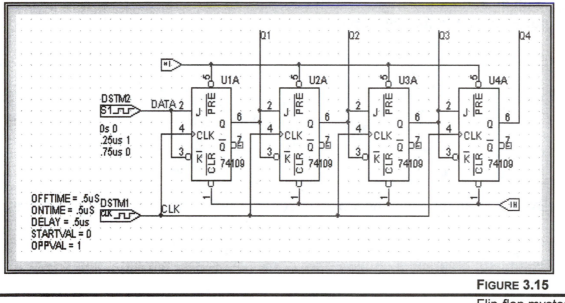

FIGURE 3.15

FIGURE 3.15

Flip-flop mystery
circuit

Questions and Problems

1. What is the difference between *level-triggered* and *edge-triggered*?

2. What makes a flip-flop *synchronous*?

3. Why is the CLR (clear) input called *asynchronous*?

4. With a 7476 JK master/slave flip-flop, what happens during the rising clock pulse and the falling clock pulse?

5. By adding an inverter, show how to construct a D-latch from a 7476 flip-flop.

6. In a function table, what does "X" mean?

7. Why is the 7475 D-latch called a "dual" latch?

Part 2

Violations and Hazards

The digital world is all about timing. In a digital circuit, it would be ideal if all events occurred at precisely the right time—not too soon or too late. Unfortunately, this is not the case. Like their analog counterparts, digital circuits have tolerances, and that makes their behavior unpredictable.

In the three chapters of Part 2, we introduce several powerful PSpice techniques for uncovering and correcting timing violations and hazards.

CHAPTER

4

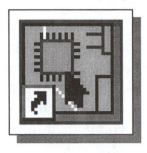

Ambiguity Hazards

Digital Worst-Case Analysis

Objectives

- *To apply worst case analysis techniques to digital circuits*
- *To identify ambiguity regions*
- *To detect and correct a variety of ambiguity hazards*

Discussion

As a production engineer, you note a puzzling situation: All your circuit boards came off the same assembly line, but 20 percent of them do not work? As with their analog counterparts, we suspect that the problem is caused by component *tolerances*. From one circuit to the next, there is a varying combination of individual component tolerances, and in 20 percent of the cases it is causing a problem.

In this chapter we will investigate the *propagation delay* tolerance parameters. *Propagation delay* is the time span between an input signal transition and the resulting output response. (In the following chapter, we will investigate the *setup*, *hold*, and *width* timing hazards.)

Ambiguity

As shown in Figure 4.1, PSpice displays all digital signals as one of six states. During normal analysis, only the *low* (0), *high* (1), and *high impedance* (Z) states are displayed. When we switch to worst-case analysis, PSpice automatically adds the *rising* (R), *falling* (F), and *unknown* (X) states. These added worst-case states reveal possible hazards to the circuit and must be addressed.

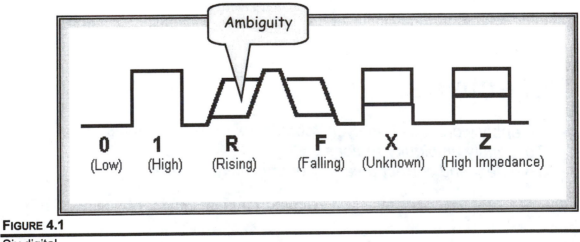

FIGURE 4.1

Six digital
states

All *R* (rising) and *F* (falling) regions are known as *ambiguity regions* because the exact time of transition is not precisely known. The major source of ambiguity is the variation in a component's *propagation delay*.

For example, Table 4.1 shows propagation delay data for the 7408 AND gate. As shown, the propagation delay is not a constant but a range of values between the *minimum* and *maximum* extremes. Most delay values will, of course, be clustered about the *typical* value. Ambiguity (uncertainty), thus, is the difference between the maximum and minimum propagation delays.

	RISING	FALLING
MINIMUM	7ns	5ns
TYPICAL	17ns	12ns
MAXIMUM	27ns	19ns
AMBIGUITY	20ns (27ns – 7ns)	14ns (19ns – 5ns)

Ambiguity.

TABLE 4.1

7408 propagation delay summary

Worst-Case Analysis

As we did with analog circuits in Volume II, we can best handle the problem of tolerance variations in digital circuits by way of *worst-case analysis*. However, instead of displaying a range of voltages, we will display *R*, *F*, and *X* ambiguity regions.

For example, based on the data of Table 4.1, Figure 4.2 shows the output ambiguity region (*R*) for just the risetime portion of the input pulse. When the input signal rises at the 40ns point, the output goes high anywhere from 7ns to 27ns later. The 20ns ambiguity region represents the interval between the earliest and the latest time that the low-to-high transition can occur.

When using worst-case analysis to uncover race (timing) problems, all possible combinations of propagation delays are generated and worst-case *R*, *F*, *X*, and timing hazards are automatically displayed.

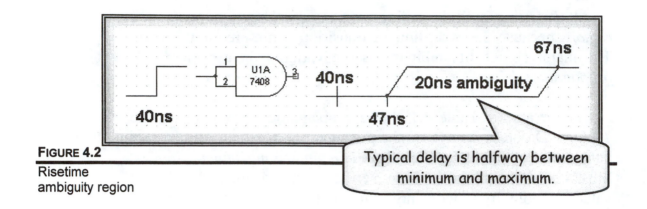

FIGURE 4.2

Risetime
ambiguity region

Simulation Practice

Activity *AMBIGUITY*

This activity tests the ambiguity characteristics of a 7408 AND gate and shows how PSpice uncovers *ambiguity hazards*.

1. Create project *tolerance* with schematic *AMBIGUITY*.

2. Draw the test circuit of Figure 4.3, and set the clock stimulus for 125MEGHz (1/80ns) as shown.

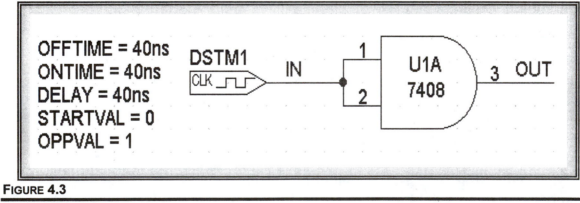

FIGURE 4.3

Propagation delay
test circuit

3.   Set the simulation profile for *Transient* from 0s to 120ns, with a step ceiling of 120ps.

4.   Set the timing mode to *Minimum* (see *Simulation Note 4.1*).

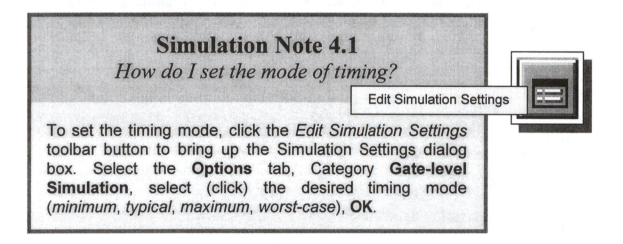

Simulation Note 4.1
How do I set the mode of timing?

Edit Simulation Settings

To set the timing mode, click the *Edit Simulation Settings* toolbar button to bring up the Simulation Settings dialog box. Select the **Options** tab, Category **Gate-level Simulation**, select (click) the desired timing mode (*minimum*, *typical*, *maximum*, *worst-case*), **OK**.

5.   Run PSpice and generate the waveforms of Figure 4.4.

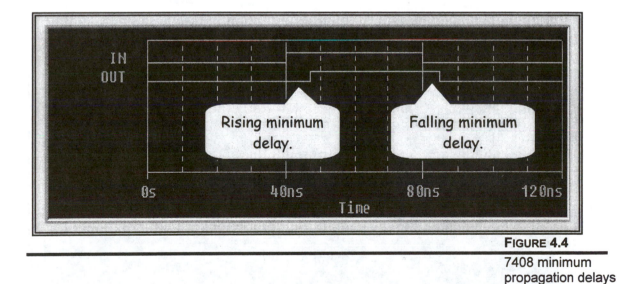

FIGURE 4.4

7408 minimum propagation delays

6.   Are the minimum rising and falling propagation delays the same as shown in Table 4.1? (Be sure to use the cursors.)

Yes   No

7. Repeat for the *typical* and *maximum* cases (as outlined in *Simulation Note 4.1*). Do they agree with Table 4.1?

   Yes   No

8. Are rising and falling (*R*, *F*) ambiguity regions generated for these minimum, typical, and maximum cases?

   Yes   No

Worst-Case Analysis

> When we switch to *Worst-Case* analysis, the system automatically generates the special *R* (rising) and *F* (falling) ambiguity regions, which reflect the difference between the minimum and maximum propagation delay times.

9. This time repeat the same process, but use *Worst-Case* analysis to generate the ambiguity curves of Figure 4.5. Do the risetime and falltime ambiguities match the predictions of Table 4.1?

   Yes   No

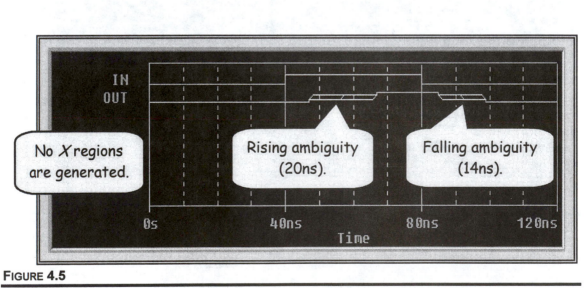

FIGURE 4.5

Worst-case run showing ambiguity regions

10. Lower the *OFFTIME*, *ONTIME*, and *DELAY* attributes of Figure 4.3 from 40ns to 10ns, and change the transient time from a maximum of 120ns down to 40ns. Rerun PSpice and generate the new worst-case curves of Figure 4.6.

   a. Was an *X* (unknown) *ambiguity hazard* region generated, and was it highlighted in red?

   Yes   No

   b. Did the beginning of the unknown *X* region correspond to the overlapping of the maximum 27ns of rising delay with the minimum of 5ns falling delay?

   Yes   No

   c. Did the end of the unknown region correspond to the 7ns of minimum rising delay from the time of the second pulse?

   Yes   No

   d. Is it possible that the NAND gate output will remain permanently HIGH after the first pulse? (*Hint*: Can the *X* region always be HIGH?)

   Yes   No

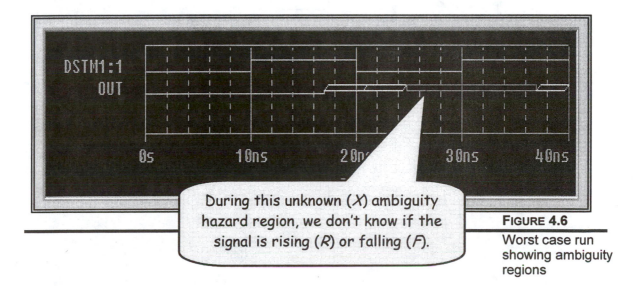

During this unknown (*X*) ambiguity hazard region, we don't know if the signal is rising (*R*) or falling (*F*).

FIGURE 4.6

Worst case run showing ambiguity regions

Activity *CONVERGENCE*

Activity *CONVERGENCE* will uncover the *ambiguity convergence hazard* of the gate array of Figure 4.7. This hazard occurs when two or more signals with overlapping ambiguity regions converge at a common point (such as a gate). For this type of hazard, the system generates a *simulation message*.

11. Add schematic *CONVERGENCE* to project *tolerance*.

12. Draw the test circuit of Figure 4.7 and set the stimulus commands as shown.

13. Set the simulation profile to *Transient* from 0 to 50ns with ceiling of 50ps and timing mode *Worst case*.

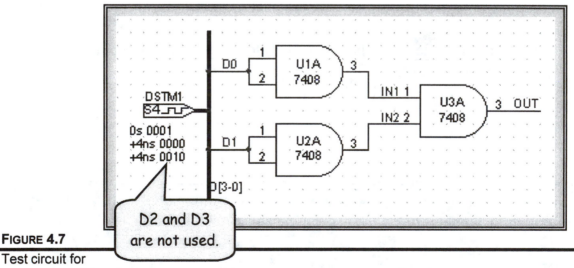

FIGURE 4.7

Test circuit for
convergence hazard

14. Reviewing the input waveform set of Figure 4.8:

   a. In the ideal case (no propagation delay) would the output (*OUT*) always be zero?

   Yes   No

b. Predict and draw the expected worst-case output waveform (*OUT*), showing the ambiguity region.

> *Hint*: When *IN2* goes high, the output becomes an ambiguous *R* a minimum 7ns later; when *IN1* goes low, the output becomes an unambiguous LOW a maximum 19ns later.

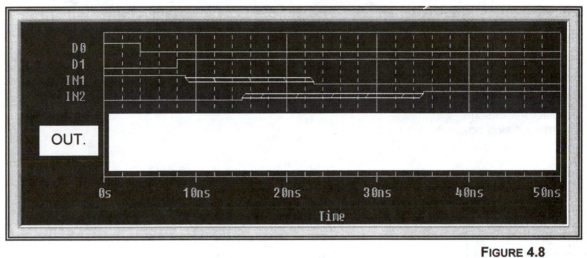

FIGURE 4.8
Convergence hazard input waveforms

15. Run the worst-case simulation. Because PSpice found a convergence violation, the Simulation Messages dialog box appears. **Yes** to view the message and bring up the Simulation Message Summary dialog box of Figure 4.9.

What type of hazard occurred? _____

When did the hazard occur? _____

What device sensed the hazard? _____

PSpice for Windows

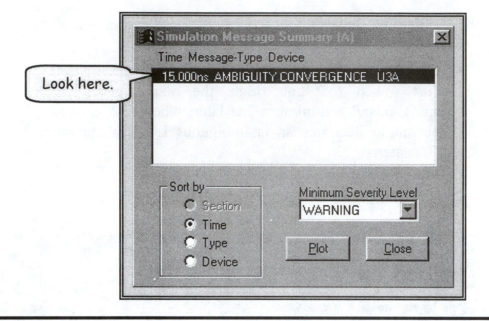

FIGURE 4.9

Simulation Message
Summary dialog box

16. Display the options available under *Minimum Severity Level.*

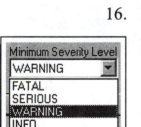

a. Was the *ambiguity convergence* hazard a *warning*?

      Yes   No

b. Were there any *serious* or *fatal* hazards detected?

      Yes   No

17. Activate the various options available under *Sort By* and note the changes. Was the first item listed in the window accordingly changed?

      Yes   No

18. Whenever a hazard occurs, we are given the option of viewing the waveforms associated with the hazard. To exercise this option, **Plot**, **Close**. By adding and rearranging curves, and expanding the X-axis, generate the curves of Figure 4.10. Were your predictions of Figure 4.8 correct?

      Yes   No

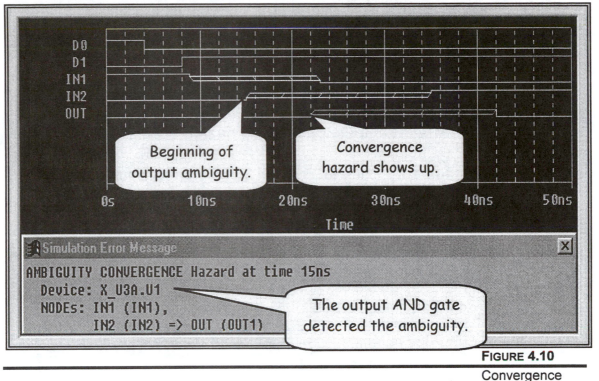

FIGURE 4.10

Convergence
hazard waveform set

19. The convergence hazard occurred at 15ns because that is when the input ambiguity regions overlapped. During this overlap period, the output waveform is uncertain. Referring to Figure 4.10:

a. What is the state of the output ambiguity region? Circle your answer.

    R   F   X

b. From Table 4.1, the minimum rising propagation delay is 7ns. Does OUT go ambiguous 7ns after the ambiguity convergence hazard (where *IN2* goes rising ambiguously)?

    Yes   No

c. From Table 4.1 the maximum falling propagation delay is 19ns. Does *OUT* go LOW 19ns after *IN1* goes unambiguously low?

    Yes   No

20. Review the Simulation Error Message (bottom of Figure 4.10):

    What device detected the ambiguity? _____

    At what input nodes did the signals converge? _____

    At what output node did the hazard occur? _____

21. Referring back to Figure 4.7, double the input signal spacing from +4ns to +8ns and rerun the worst-case analysis.

    a. Was the ambiguity convergence hazard eliminated? (Was the input ambiguity region overlap reduced to below the minimum rising delay of 7ns?)

    Yes   No

    b. Is the output signal zero at all times (as predicted for ideal gates in step 14)?

    Yes   No

Activity *CUMULATIVE*

Activity *CUMULATIVE* demonstrates the *cumulative ambiguity* hazard of Figure 4.11. This occurs when signals propagate through layers of gates and ambiguity accumulates. When the rising-edge ambiguity overlaps the falling-edge ambiguity, an unknown ("X") output region is created.

---

22. Add schematic *CUMULATIVE* to project *ambiguity*.

23. Draw the test circuit of Figure 4.11. Set the simulation profile to *Transient* from 0 to 200ns, step ceiling of 200ps, timing analysis *Worst Case*.

24. Run PSpice and generate the hazard curves of Figure 4.12. (*Hint*: Add and rearrange curves as necessary, and expand the X-axis.)

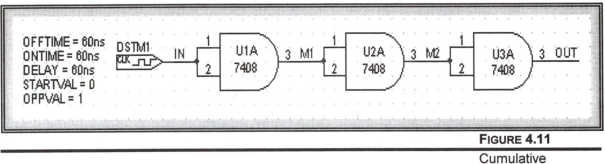

FIGURE 4.11

Cumulative
ambiguity test circuit

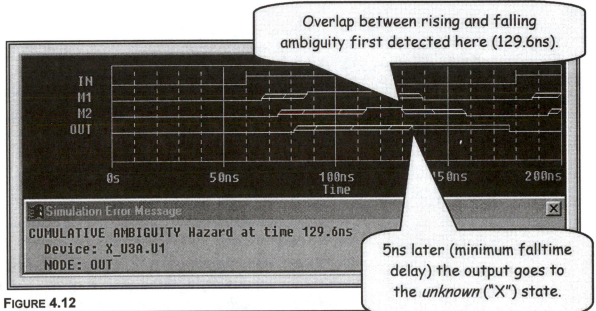

FIGURE 4.12

Cumulative
ambiguity
waveform set

25. Analyze the curves and answer the following:

   a. As the input waveform propagates through the gates, does the ambiguity accumulate (Do the ambiguity regions grow wider?), and is the regular (unambiguous) portion of the signal gradually squeezed out?

   Yes   No

b. Does the output signal (OUT) show an unknown (X) region where the signal could either be rising or falling?

Yes   No

c. Does the X region of the output signal (*OUT*) first appear 5ns after the overlapping ambiguity is detected?

Yes   No

26. Increase the input pulse width from +60ns to +80ns. Did the cumulative hazard disappear?

Yes   No

Advanced Activities

27. Increase the clock speed of activity *CONVERGENCE* still further (beyond that of Figure 4.7) and explain all results.

28. Perform a worst-case analysis on the oscillator of Figure 4.13 (from Chapter 1) and generate the output graph of Figure 4.14.

> Because of the continuous accumulation of ambiguity in the feedback loop, the output signal quickly reaches the X (unknown) state and incorrectly ceases to operate. *We conclude that worst-case analysis cannot be used in circuits with asynchronous feedback.*
>
> To overcome the problem, set all active devices in the feedback loop (the 7414s) to typical (TYP) by changing model parameter MNTYMXDLY from 4 (*Worst Case*) to 2 (*Typical*). All other devices, if any, will continue to operate in the worst-case mode. Does the circuit now operate properly?

29. Increase Figure 4.11 to four gates and perform a complete ambiguity analysis. Based on your results, predict and test the outcome on five gates in series.

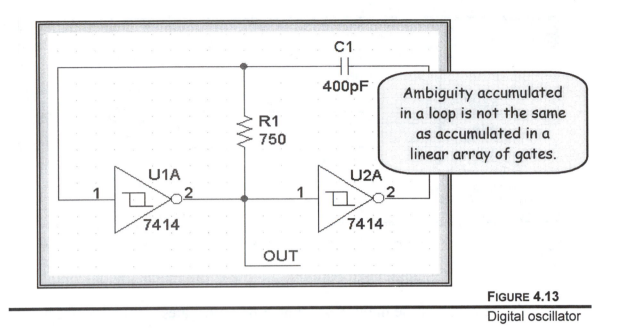

FIGURE 4.13

Digital oscillator

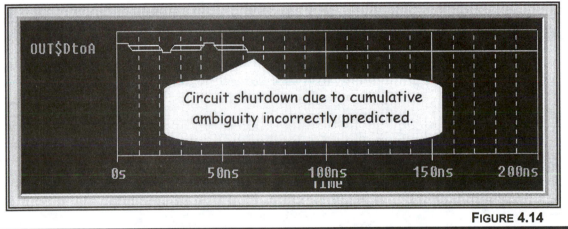

FIGURE 4.14

Ambiguity build-up
in oscillator

Exercises

1. Referring to Figure 4.15, predict the highest safe frequency of operation (no "X" states in the output signal). Verify your predictions using PSpice (worst- case analysis).

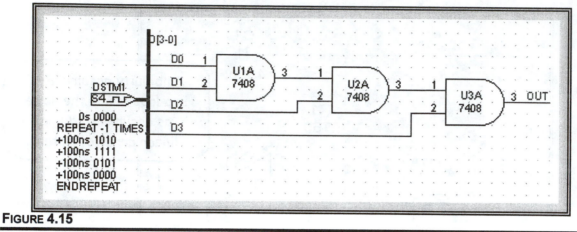

FIGURE 4.15

Exercise circuit

2. Determine the largest number of AND gates that can be safely placed end to end for a 100MEGHz input square wave signal.

Questions and Problems

1. Typically, propagation delays for gates fall into which of the following time spans?

   a. ps
   b. ns
   c. μs
   d. ms

2. The input waveform to 10 AND gates in series goes high at time t = 0. What is the very earliest time that the output will go high? The very latest?

3. Rising and falling ambiguity times are the same.

          True   False

4. If at least one input to an AND gate is always low, does this guarantee that the output will also always be low—regardless of gate timing?

    Yes   No

5. The input waveform shown below has input rising and falling ambiguities of 1ns. After passing through the 7404, which adds additional ambiguities, draw the output waveform (including ambiguity regions).

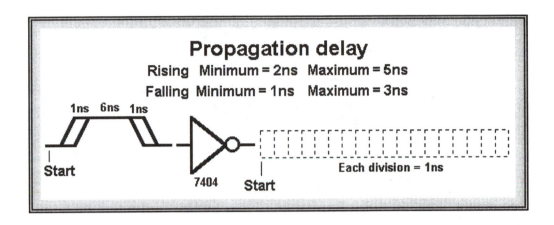

6. Why is it unrealistic to perform a worst-case analysis on a feedback loop?

Timing Hazards

Setup, Hold, and Width

Objectives

- *To identify and correct a variety of timing violations*
- *To display multiple hazards*

Discussion

We live in a world in which timing plays a vital role. We are told to arrive at the airport one hour before departure. We are told to press a button for three seconds to reset our digital watches. Timing requirements also are found throughout digital and computer circuits.

In this chapter we will investigate the following timing specifications.

- *Setup* — the time span during which a signal must be held steady *before* an action is taken.

- *Hold* — the time span during which a signal must be held steady *after* an action is taken.

- *Width* — the required width of a clock pulse.

As before, the best way to locate and eliminate these timing hazards is to enter the worst-case mode and let PSpice perform an exhaustive search of the waveform set.

Simulation Practice

Activity *SHW*

Activity *SHW* uses Figure 5.1 to test the *setup*, *hold*, and *width* hazards. These hazards commonly occur in circuits that are clocked, such as flip-flops. If the width of a signal is too low, a width hazard is predicted; if not stable for a sufficient time before or after clock occurs, a setup or hold hazard is predicted.

1. Add schematic *SHW* to project *timinghazard*.

2. Draw the circuit of Figure 5.1 and set the timing commands as shown.

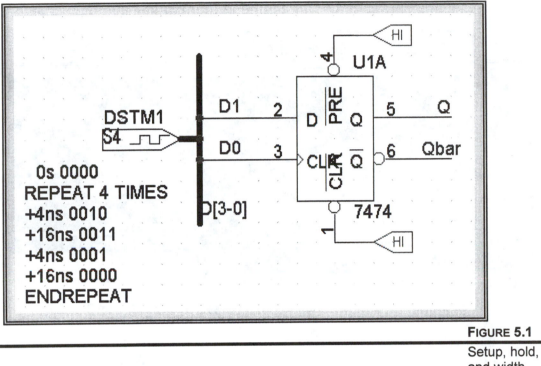

FIGURE 5.1

Setup, hold, and width test circuit

3. Set the simulation profile for *Transient* from 0 to 100ns, step-ceiling 100ps, and *Worst-Case* timing mode, and initialize all flip-flops to 0.

4. Run PSpice and **Yes** (8 messages occurred) to bring up the Simulation Message Summary dialog box. Note that the first three hazards are SETUP (20ns), HOLD (24ns), and WIDTH (40ns).

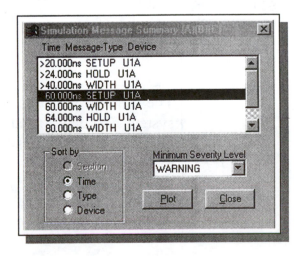

5. To create three separate windows corresponding to these hazards, **Plot, Plot, Plot, Close**. (Note that tabs A, B, and C identify the windows.)

PSpice for Windows

Setup

6. First, select the Setup window of Figure 5.2. (Click the A tab.)

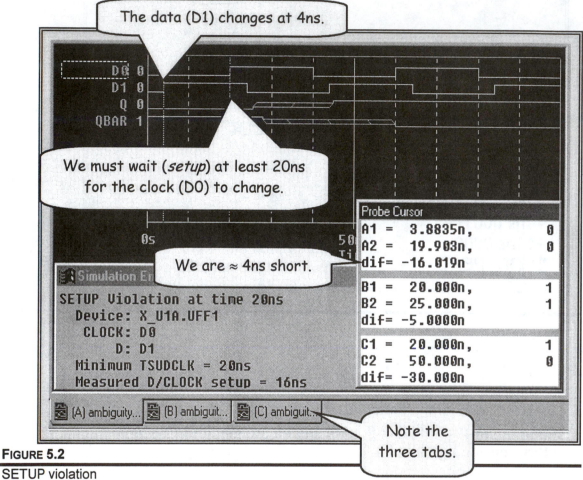

FIGURE 5.2

SETUP violation
waveform set

The SETUP violation occurred at 20ns because that is when the system first determined that the time between a data change (D1 at 4ns) and a clock transition (D0 at 20ns) was below the minimum required 20ns.

Note the resulting ambiguity in the output waveforms. The output is predicted to eventually go high at 45ns.

7.   With the help of the Simulation Error Message box, record each of the following:

   Measured D/CLOCK *Setup* = _____

   Minimum *Setup* (TSUDCLK) = _____

   Additional *Setup* (minimum – measured) = _____

8.   In your own words, what caused the SETUP hazard?

Hold

9.   Next, select the Hold window (the B tab) of Figure 5.3.

> The HOLD violation occurred at 24ns because that is when the system first determined that the data input (D1 at 24ns) changed too quickly after the clock (D0 at 20ns) went active high.

10.   With the help of the Simulation Error Message box, determine each of the following:

   Measured D/CLOCK *Hold* = _____

   Minimum required *Hold* (THDCLK) = _____

   Additional *Hold* (minimum – measured) = _____

11.   In your own words, what caused the HOLD hazard?

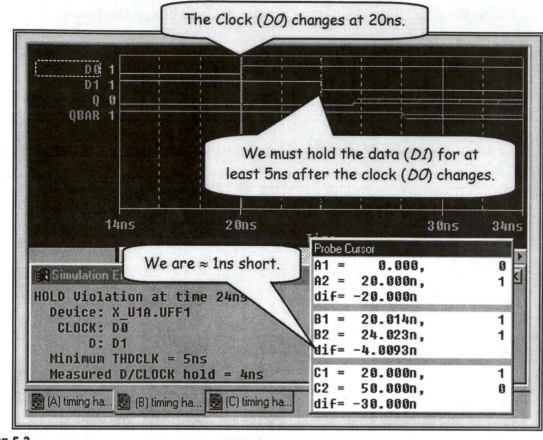

FIGURE 5.3

HOLD violation
waveform set

Width

12. Finally, select the Width window (C tab) of Figure 5.4.

> The WIDTH hazard occurred at 40ns because that is when the system first determined that the width of the clock signal (D0) was too small. Record each of the following:

Measured CLOCK pulse *Width* = _____

Minimum *Width* (TWCLKH) = _____

Additional *Width* (minimum – measured) = _____

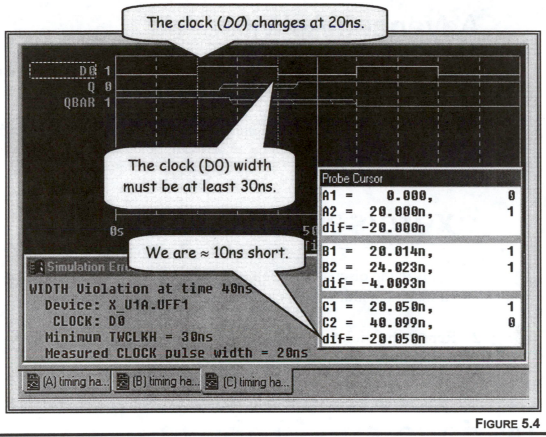

FIGURE 5.4

WIDTH violation
waveform set

13. In your own words, what caused the WIDTH violation?

14. Double the command times in Figure 5.1 from 4ns/16ns to
8ns/32ns. Were all the violations eliminated? (If you wish,
use the cursors to verify that the *setup*, *hold*, and *width* times
are at or above the minimums required.)

           Yes   No

Advanced Activities

15. Feed the Q and Qbar outputs of the flip-flop (Figure 5.1) to an AND gate and explain the resulting waveform.

16. Analyze any of the additional five violations (see step 4) and summarize your results.

Exercises

1. Analyze the discrete flip-flop of Chapter 3 (Figure 3.2) for Setup, Hold, and Width violations.

2. Repeat the violation analysis of this chapter for the 7475 level-triggered flip-flop.

Questions and Problems

1. Why would a Setup violation normally occur before the corresponding Hold violation?

2. Why can't the data and clock to a flip-flop (such as the 7474) occur (change) at precisely the same time?

3. Referring to Figure 5.2, why does the Q output stay high after the first clock pulse?

4. Given the following waveform set, a Setup violation occurred at 20ns because the data must be stable at least 30ns before the clock changes. Why is it possible that a design engineer might decide to ignore this violation?

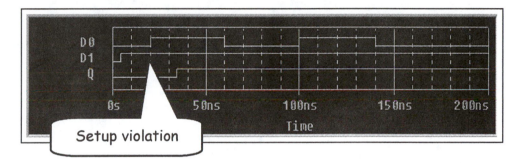

5. Is it reasonable to assume that special flip-flops exist on the market with "tighter" (lower) timing specifications?

CHAPTER

6

Critical
Hazards

Persistence

Objective

- *To identify and correct a class of hazards that usually cannot be ignored*

Discussion

All the ambiguity and timing hazards of the previous two chapters were warnings. As such, they may or may not cause a serious problem. They are simply events that the designer must examine in order to determine if a change in circuit design is warranted. As an example, consider the D-Latch circuit of Figure 6.1.

When the circuit is processed under PSpice, the waveform set of Figure 6.2 shows that a setup warning was properly identified at 20ns. This is because the data line transition (*D1*) occurred only 10ns before the active high clock pulse (*D0*), and a minimum of 20ns is required.

However, this probably will not cause a problem in this case because a second clock pulse comes along at 80ns and properly latches the data. (Note that the ambiguity in the Q signal occurs only after the first clock pulse.) Therefore, the design engineer may very well decide to ignore this particular setup warning.

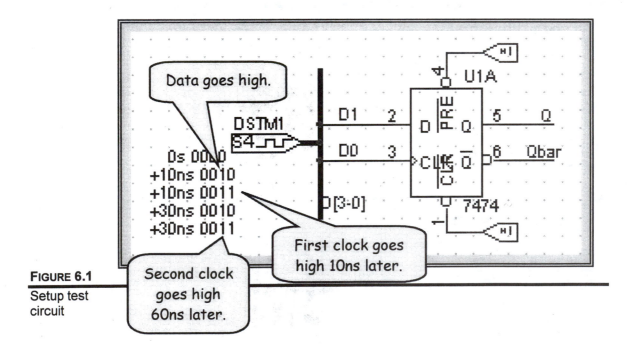

FIGURE 6.1

Setup test circuit

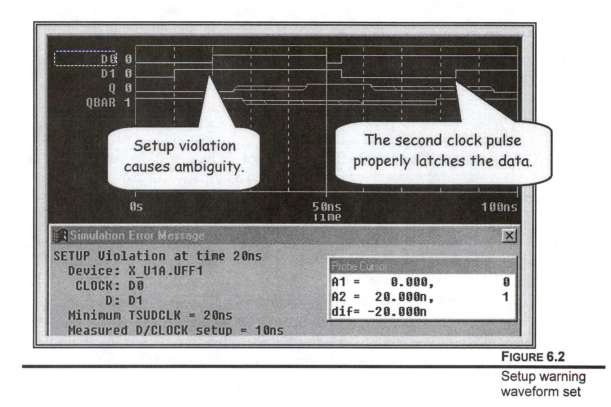

FIGURE 6.2

Setup warning
waveform set

Persistent Hazard

A *persistent hazard*, on the other hand, is either a timing violation or hazard that can cause an incorrect state to be latched into an internal circuit (such as a flip-flop) or one that is passed on to a primary circuit output.

These more serious persistent hazards, which usually cannot be ignored, are a major subject of this chapter.

Design Methodology

Because of violations and hazards, the development of a digital circuit is a two-phase process: *design* and *verification*.

- During *design*, all tolerances are set to typical and we concentrate on the *state response* of the circuit.

- During *verification*, all tolerances are set to worst-case and we concentrate on violations and hazards. Starting at a hazard point, we work backward until the cause of the hazard is uncovered and corrected.

PSpice for Windows

Simulation Practice

Activity *PORT*

Activity *PORT* uses the circuit of Figure 6.3 to show how a persistent hazard is triggered when a timing violation is passed on to another circuit by way of a port interface.

1.   Add schematic *PORT* to project *persistenthazard*.

2.   Draw the hazard test circuit of Figure 6.3, and set the stimulus commands as shown.

To place the external port, click the *Place Port* toolbar button, library *CAPSYM*, select *PORTBOTH-L* (*in and out, left connect*), and place as shown.

> A warning hazard becomes a persistent hazard if it is passed on to other circuits by way of a port.

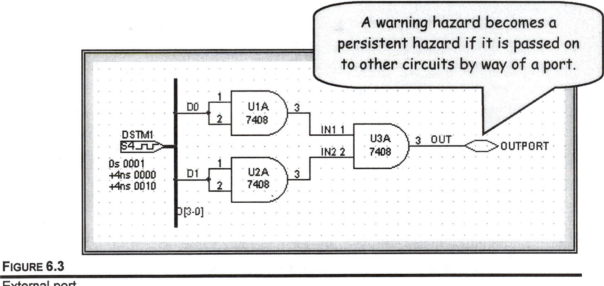

FIGURE 6.3

External port
added

3.   Set the simulation profile to *Transient*, from 0s to 100ns, step ceiling of 100ps, *Worst-Case* mode.

4. Run a *worst-case* simulation, **Yes** (to view messages). Note that a *SERIOUS* hazard is indicated by the Simulation Message Summary dialog box. **Plot, Close** to generate the hazard waveform set of Figure 6.4. As shown, two plots are now generated for these more serious hazards:

   ▪ The upper plot displays the persistent hazard at 22ns due to the propagation of a timing hazard to an output port.

   ▪ The lower plot displays the *cause* of the persistent hazard—a convergence hazard at 15ns. (Does 22ns – 15ns = 7ns, the *minimum risetime delay* for a 7408?)

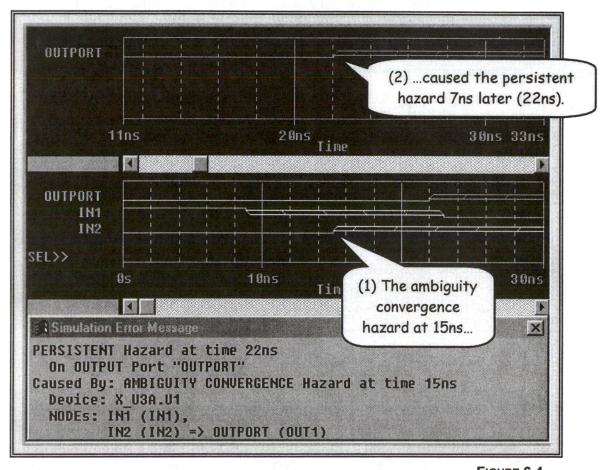

FIGURE 6.4

Persistent hazard
and cause

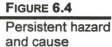

5.   Increase the timing from 4ns to 8ns. Was the hazard eliminated, and did the output of the circuit (*P1*) remain low?

Yes   No

Activity *LATCH*

Activity *LATCH* uses the circuit of Figure 6.5 to show how a persistent hazard is created when a warning hazard is latched.

---

6.   Add schematic *LATCH* to project *persistenthazard*.

7.   Draw the circuit of Figure 6.5 and set the stimulus commands as shown.

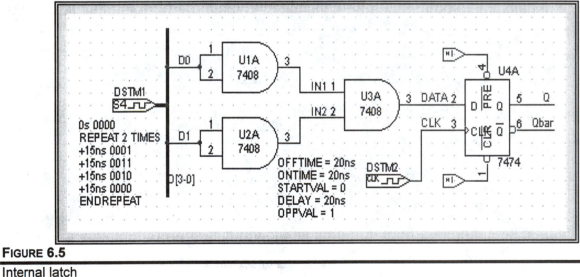

FIGURE 6.5

Internal latch
test circuit

8.   Run the *worst-case* simulation. Were five messages generated, one *SERIOUS* and four *WARNING*?

Yes   No

9.   Plot the persistent (SERIOUS) hazard, generating the waveform set of Figure 6.6.

- The upper plot displays a persistent hazard at 60ns because an ambiguity state (R) was present at the data input (Data) to the flip-flop when the clock (CLK) went active high.

  As a result, an unknown value was latched into the flip-flop, and its output (Q) went indeterminate (state X) at 67ns. This could cause a serious malfunction in the overall circuit.

- The lower plot shows us that the cause of the persistent hazard was the ambiguity convergence hazard at the output of the AND gate array at 49.8ns.

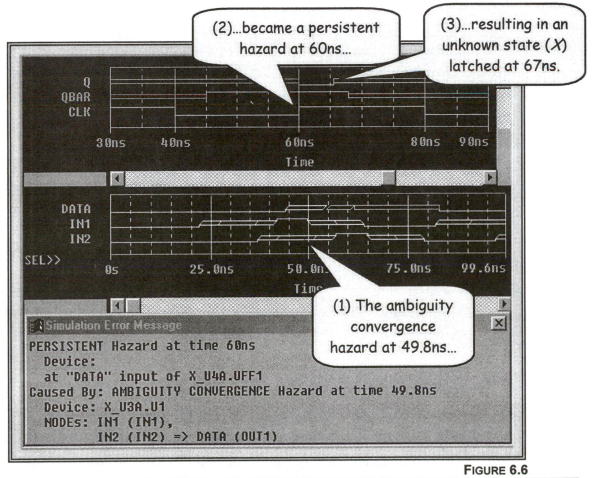

FIGURE 6.6

Persistent hazard and cause

10. Increase all DSTM1 times from 15ns to 60ns and all DSTM2 times from 20ns to 40ns and rerun the simulation. Were all hazards eliminated and did the *Data* and *Q* lines remain low?

Yes   No

Advanced Activities

11. Using worst-case analysis, determine the maximum frequency of operation of the D-latch–based ripple counter of Figure 6.7. (As the frequency is increased, what is the first hazard that appears?)

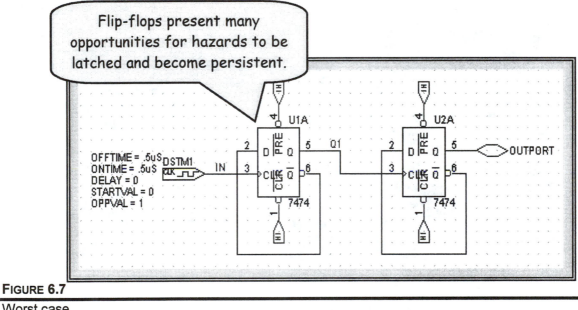

FIGURE 6.7

Worst case
test circuit

12. Refer back to the JK flip-flop test circuit of Figure 3.11. Add an in/out port to the output, decrease all times to the nanosecond range (change μ to n), and perform a complete worst-case analysis.

Explain how the asynchronous inputs (*PRE* and *CLR*) are involved in the production of warnings and hazards.

Exercises

1. Perform the two-step design methodology outlined in the discussion on the circuit of Figure 6.8 (reproduced from Figure 6.5.)

   ▪ Set the timing state to *typical*, run the simulation, and generate a waveform set.   Do the waveforms indicate that the state response of the circuit is correct?

   ▪ Switch to worst case mode and generate a hazard waveform set. Report how you corrected the hazard.

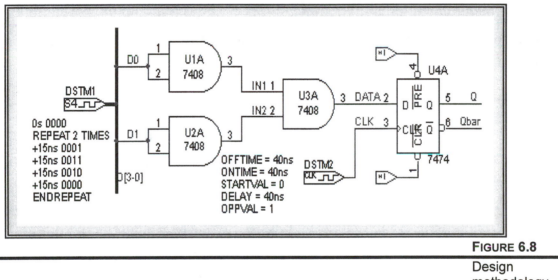

FIGURE 6.8

Design methodology test circuit

2. Refer back to the parity test circuit of Figure 3.14.   Add an *in/out* port to the output, decrease all times to the nanosecond range (change μ to n), and perform a worst-case analysis.

   a. Were both SERIOUS and WARNING hazards detected?

   b. How were they corrected?

Questions and Problems

1. What two conditions does PSpice look for when tagging a hazard as *persistent*?

2. Can some WARNING hazards be ignored? Why?

3. Reviewing Figure 6.2, what is the setup time of the data (D1) for the *second* clock pulse (D0)?

4. What two basic steps should every design engineer perform on a **digital** circuit?

5. Does PSpice display both the *cause* and *effect* of a persistent hazard?

6. Referring to Figure 6.4, why does the persistent hazard show up 7ns after the ambiguity convergence warning?

Part 3

Counters, Shift Registers, Coders, and Timers

A marching band is a group of musicians, all highly trained to step in unison to a single drummer. Digital circuits, especially those associated with computers, must follow a similar pattern.

In the four chapters of Part 3, we learn about circuits that create and manipulate groups of bits as they march to and fro through the digital system. We also learn about a timing circuit which helps to regulate the basic drumbeat that drives the circuit forward.

7

Counters

Synchronous Versus Asynchronous

Objectives

- *To compare synchronous versus asynchronous digital circuitry*
- *To define the modulus of a counter*
- *To design and test a variety of counters*

Discussion

A digital counter is a device that keeps track of clock cycles (counts). Counters are fashioned from a series of flip-flops strung together in sequence, with one flip-flop feeding the next.

The number of different binary states defines the *modulus* (MOD) of the counter. For example, if a counter counts from 0 to 3, it is a MOD-4 counter. A MOD-4 counter must be constructed from two flip-flops.

Synchronous Versus Asynchronous

Digital counters are classified as either *synchronous* or *asynchronous*.

- If the output of one flip-flop is used to clock the next flip-flop, as shown in the MOD-4 counter of Figure 7.1(a), the counter is asynchronous. Asynchronous counters are also called *ripple counters* because the clock signal ripples from one flip-flop to the next.

- If all the flip-flops are clocked at the same time, as shown in the MOD-4 counter of Figure 7.1(b), the counter is synchronous. Synchronous counters are much faster and more precise than asynchronous counters because they do not have to wait for the signal to propagate through the flip-flops.

Odd Modulus Counters

The natural modulus of a counter is 2^N, where N is the number of flip-flops. For example, we see from Figure 7.1 that the natural count of a two-flip-flop counter is 4 (0 to 3).

However, suppose we desired an unnatural modulus—such as 3 (0 to 2). As shown by Figure 7.2, we use feedback and feedforward to properly program the flip-flops.

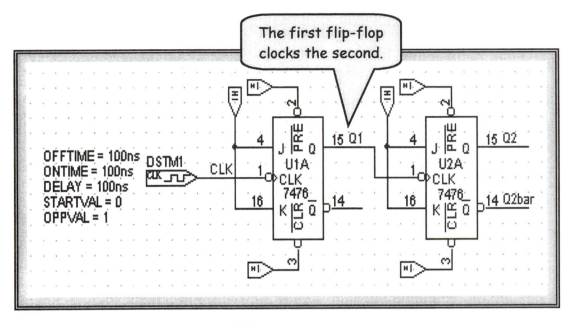

(a) Ripple counter

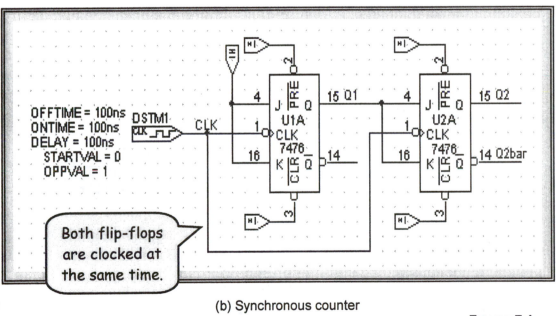

(b) Synchronous counter

FIGURE 7.1

Two-stage MOD-4
binary counters
(a) Ripple counter
(b) Synchronous counter

PSpice for Windows

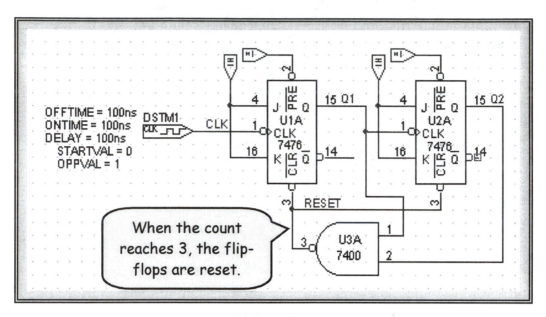

(a) Asynchronous

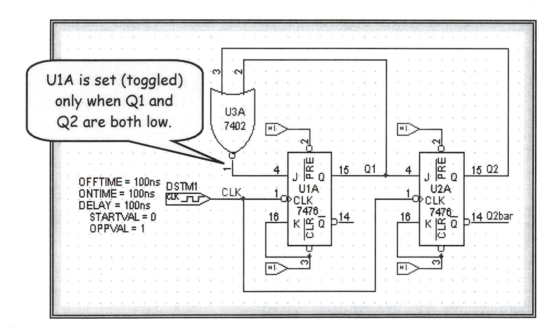

(b) Synchronous

FIGURE 7.2

MOD 3 counters
(a) Asynchronous
(b) Synchronous

PSpice for Windows

- Refer to the asynchronous counter of Figure 7.2(a). When the count reaches three, an active-low reset pulse clears both flip-flops back to zero. The brief period of time that the three-state exists generates a glitch—which can be a problem at high frequencies.

- Refer to the synchronous counter of Figure 7.2(b). The present state of the flip-flops programs the next state. For example, with the 0, 1, 2 count sequence below, Q2 toggles high whenever Q1 is high, and Q1 toggles high only when both Q1 and Q2 are low. Because there is no asynchronous feedback, no glitches are produced.

Count	Q2	Q1
0	0	0
1	0	1
2	1	0

Integrated Circuit Counters

Fortunately, a wide variety of integrated circuit counters are available to simplify circuit design. For 4-bit counters, the most popular modulus numbers are 10 and 16. The MOD 10 (BCD) counters use internal feedback to reset the count upon reaching 10 (binary 1010).

After preliminary work with the simple discrete MOD-3 and MOD-4 counters, we will work primarily with the 7490 MOD-10 binary ripple counter and the 74190 MOD 10 synchronous counter.

Simulation Practice

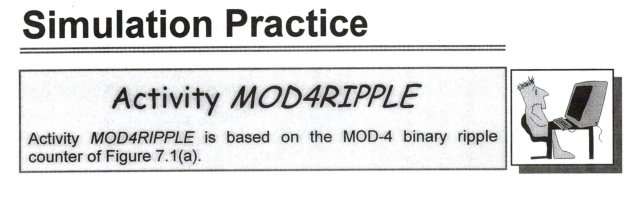

Activity *MOD4RIPPLE*

Activity *MOD4RIPPLE* is based on the MOD-4 binary ripple counter of Figure 7.1(a).

1. Create project *counter*, with schematic *MOD4RIPPLE*.

2. Draw the MOD-4 (2-bit) binary ripple counter circuit of Figure 7.1(a), and set the attributes as shown.

3. On the graph of Figure 7.3, predict (draw) the Q1 and Q2 waveforms in the spaces provided.

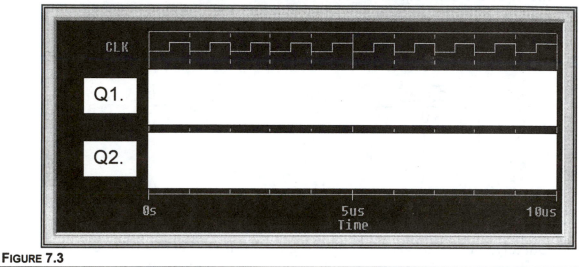

FIGURE 7.3

MOD-4 ripple
counter waveform

4. Based on your predictions, what is the frequency relationship between CLK and Q2.

   CLK frequency = _____ × Q2 frequency

5. Set the simulation profile to *Transient* from 0 to 10µs, step ceiling 10ns, *typical* timing mode. Be sure to initialize all flip-flops to the zero state by **Options, Gate-level Simulation,** *Initialize all Flip-flops to* **0**.

6. Run PSpice and generate the *CLOCK, Q1,* and *Q2* waveforms. Did the results match your predictions? (Did the count go from 0 to 3, and was *CLK* four times the frequency of *Q2*?)

   Yes   No

7. Switch to *worst-case* timing and repeat.   Did the ambiguity regions widen from Q1 to Q2?

   Yes   No

Activity *MOD4SYNC*

Activity *MOD4SYNC* is based on the MOD-4 binary synchronous counter of Figure 7.1(b).

8. Add schematic *MOD4SYNC* to project *counter*.

9. Repeat steps 1 through 7 for the synchronous counter of Figure 7.1(b).   Were the results the same, except that the worst-case ambiguity regions of Q1 and Q2 remained constant (did not widen)?

   Yes   No

Activity *MOD3RIPPLE*

Activity *MOD3RIPPLE* is based on the MOD-3 binary ripple counter of Figure 7.2(a).

10. Add schematic *MOD3RIPPLE* to project *counter*.

11. Draw schematic *MOD3RIPPLE* of Figure 7.2(a).

12. On the graph of Figure 7.4, draw the expected output waveform from Q1 and Q2.

13. Set the simulation profile to *Transient* from 0 to 2µs, step ceiling 2ns, *typical* timing.

14. Generate the *CLK, Q1, Q2,* and *RESET* waveforms for the MOD-3 ripple counter of Figure 7.2(a).

a. Were the waveforms as predicted? (Did the count go from 0 to 2)?

        Yes   No

b. Was a short RESET glitch observed in Q1 when the count went briefly to 3?

        Yes   No

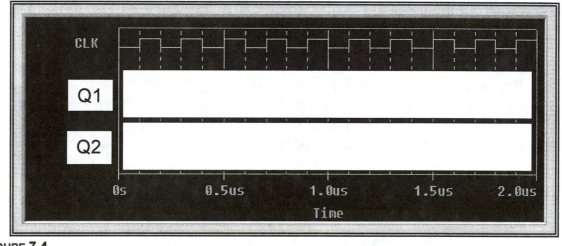

FIGURE 7.4

MOD3RIPPLE
waveforms

Activity *MOD3SYNC*

Activity *MOD3SYNC* is based on the MOD-3 binary synchronous counter of Figure 7.2(b).

15.   Add schematic *MOD3SYNC* to project *counter*.

16.   Draw the circuit of Figure 7.2(b).

17.   Set the simulation profile for *Transient* from 0 to 2μs, step ceiling 2ns, *typical* timing.

18. Run PSpice and display the *CLK*, *Q1*, and *Q2* waveforms. Were the results the same, except for the absence of any reset glitch after the 2 count?

    Yes   No

Activity *7490A*

Activity *7490A* is based on the popular integrated MOD-10 (decade) ripple counter of Figure 7.5.

The 7490A cascades a MOD-2 (1 flip-flop) and a MOD-5 (3-flip-flop) ripple counter to create a MOD-10 counter. The MOD-5 counter uses feedback to reduce its natural modulus from 8 to 5. For user flexibility, the connection between the MOD-2 and MOD-5 counters is done externally. Because of internal ripple delays, decoded outputs are subject to spikes.

19. Add schematic *7490A* to project *counter*.

20. Draw the asynchronous counter/decoder combination circuit of Figure 7.5, and set all attributes as shown. (Is the clock speed set higher for this integrated-circuit counter?)

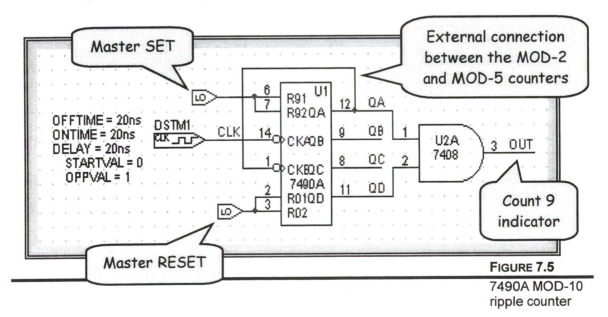

FIGURE 7.5

7490A MOD-10 ripple counter

21. Set the simulation profile to *Transient* from 0 to 6μs, step ceiling 6ns, *typical* timing.

22. Generate the graph of Figure 7.6.

    a. Did the count go from 0 to 9, and did the output decoder detect the 9 count properly?

        Yes   No

    b. Is there evidence of ripple delay?

        Yes   No

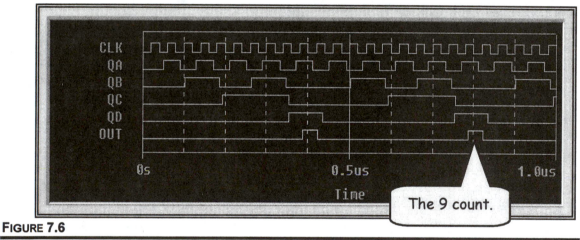

FIGURE 7.6

7490A waveforms

23. Switch to *worst-case* analysis and repeat. (**No**, when asked to view message summary.)

    a. Was there a glitch observed after each *OUT* pulse?

        Yes   No

    b. Did the ambiguity generally increase as we went from QA to QC?

        Yes   No

Activity *74160*

Activity *74160* is based on the integrated MOD-10 synchronous counter of Figure 7.7.

The 74160 synchronous counter uses *carry-look-ahead* to clock all flip-flops simultaneously. Therefore, the decoded outputs are not subject to spikes and can be used for clocks or strobes.

24.   Add schematic *74160* to project *counter*.

25.   Draw the synchronous counter circuit of Figure 7.7.

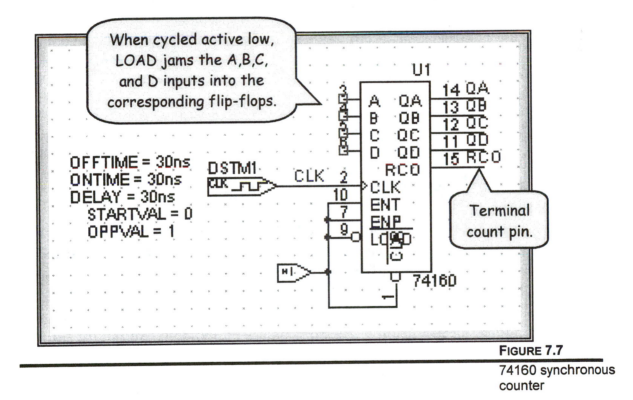

FIGURE 7.7

74160 synchronous counter

26.   Set the simulation profile to *Transient*, from 0 to .8µs, step ceiling of 1ns, *typical* timing mode.

27. Run PSpice and generate a set of waveforms (*QA*, *QB*, *QC*, *QD*, and *RCO*).

    a. Does the counter cycle properly from 0 to 9, and does RCO give the terminal count?

    　　　Yes　　　　　　No

    b. Is there any evidence of delay from QA to QD?

    　　　Yes　　　　　　No

28. Switch to *worst-case* analysis and repeat.

    a. Were glitches observed anywhere in the RCO (terminal count) signal?

    　　　Yes　　　　　　No

    b. Did the ambiguity generally stay constant as we went from QA to QD?

    　　　Yes　　　　　　No

Advanced Activities

29. Add a second cascaded 74160 to the circuit of Figure 7.7 and determine the counter's modulus. (*D* output of the first to the *CLK* input of the second.)

30. What might the circuit of Figure 7.8 be used for? What is the purpose of each component? (*Hint*: Display the output waveform from 0 to 5 seconds.)

Exercises

1. Using a design of your choice, construct a MOD-5 counter.

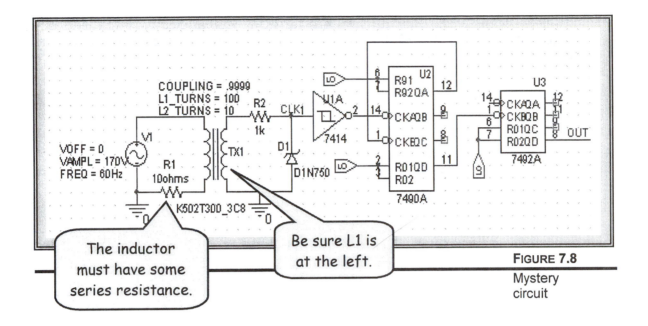

FIGURE 7.8

Mystery
circuit

2. Using the 74190 synchronous decade counter, design a circuit that counts from 0 to 9 and then from 9 to 0 repeatedly. (*Hint*: Use the terminal count to toggle a flip-flop connected to the direction input.)

3. Using the 74160 synchronous counter of Figure 7.7, design a MOD-7 (2 to 8) counter.

Questions and Problems

1. How many flip-flops would be required to design a counter that counts from 0 to 63?

2. What is the difference between *synchronous* and *asynchronous* counters?

3. Because of the minimum width requirements of the clock, the maximum frequency for both the 7490A asynchronous and 74060 synchronous counters are the same. What then is the advantage of the synchronous counter?

4. Explain how the MOD-3 synchronous counter of Figure 7.2(b) goes from the "2" state back to the "0" state.

5. Why does ambiguity increase in a ripple counter as we go from QA to AC?

6. When using the 74160 synchronous decade counter of Figure 7.7, why can a count begin from any value?

7. Referring to the mystery circuit of Figure 7.8, what is the purpose of the 7414 Schmidt trigger?

8

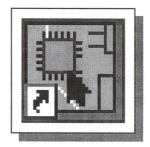

Shift
Registers

Discrete Versus Integrated

Objectives

- *To design and test a parallel-to-series shift register using both discrete and integrated circuits*
- *To design and test a Johnson counter*

Discussion

Shift Register

Shift registers are similar to counters because they are composed of a string of flip-flops. The primary function of a shift register is the movement of serial data left and right. As shown by Figure 8.1, a shift register can be composed of a linear array of edge-triggered D flip-flops. The waveform set of Figure 8.2 shows that the logic one input at the left was shifted one position through the register on every clock cycle.

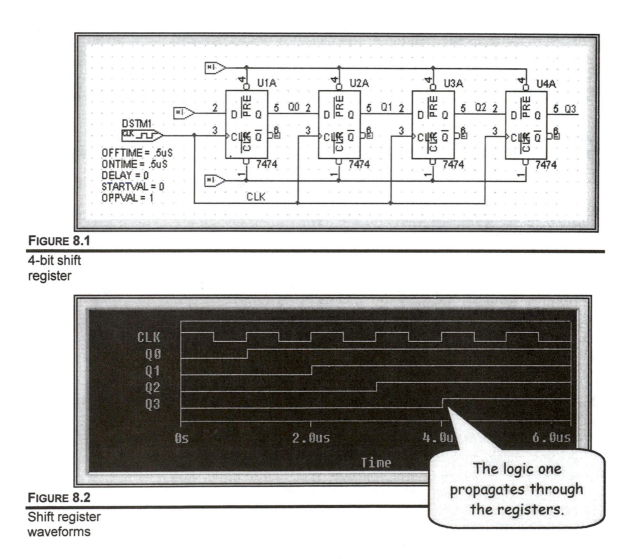

FIGURE 8.1

4-bit shift register

FIGURE 8.2

Shift register waveforms

PSpice for Windows

Besides simple shift-left and shift-right operations on serial data, shift registers process parallel data as well.   Following are the most common shift register functions.

- Parallel-to-serial
- Serial-to-parallel
- Serial-to-serial
- Parallel-to-parallel
- Shift counters

As with the counters of the previous chapter, these various operations can be performed using either discrete or integrated circuits.

Simulation Practice

Activity *P2SD*

Activity *P2SD* (Parallel-to-serial discrete) is based on the 4-bit parallel-to-serial shift register of Figure 8.3. Four bits are loaded in parallel and walked out in serial.

1. Create project *shift* and add schematic *P2SD*.

2. Draw the 4-bit parallel-to-serial shift register of Figure 8.3.

3. Set the simulation profile to *Transient*, from 0 to 2µs, step ceiling 2ns, *typical* timing mode, flip-flop initialization to zero.

4. In the space provided in Figure 8.4, predict the *Q* waveforms, where *Q0* is the final serial output. (Note that zeros are walked in from the left as the data is transmitted from the right.)

5. Generate the waveforms using PSpice, compare to your predictions, and make any necessary corrections. Place an arrow at the point where parallel data is loaded.

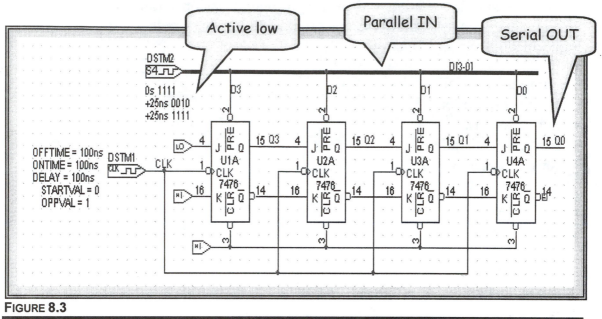

FIGURE 8.3

4-bit parallel-to-
serial shift register

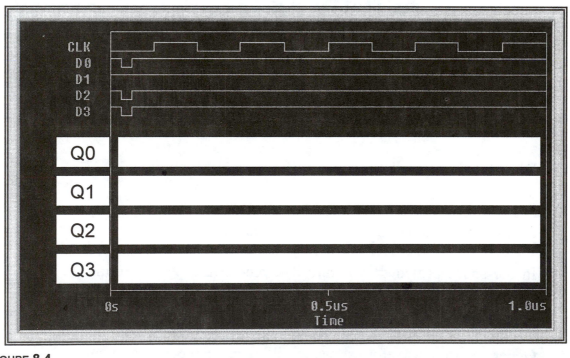

FIGURE 8.4

Shift register
waveforms

PSpice for Windows

Activity *P2SI*

Activity *P2SI* (parallel-to-serial-integrated) performs the same operation as the previous schematic but uses the single 74194 universal shift register of Figure 8.5.

The 74194 is universal because it can be programmed by the S0 and S1 pins for input or output in serial or parallel according to the table shown below.

Operating Mode	S1	S0
Hold	0	0
Shift left	1	0
Shift right	0	1
Parallel load	1	1

6. Add schematic *P2SI* to project *shift*.

7. Draw the circuit of Figure 8.5.

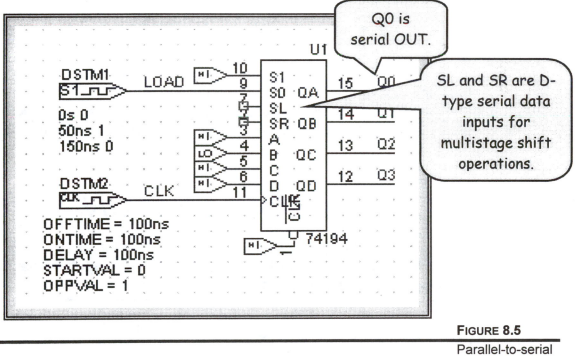

FIGURE 8.5

Parallel-to-serial conversion using The 74194

PSpice for Windows

8. On the graph of Figure 8.6, predict (draw) the output waveforms listed.

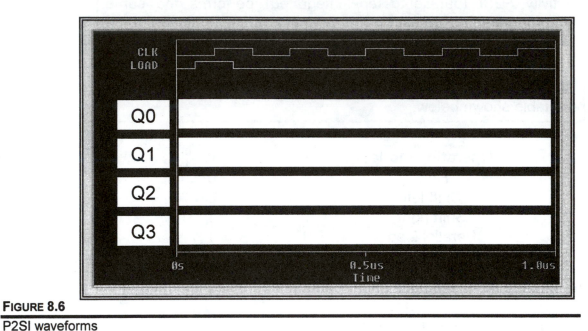

FIGURE 8.6

P2SI waveforms

9. Set the simulation profile for *Transient* from 0 to 1µs, step ceiling 1ns, *typical* timing, flip-flops to zero.

10. Run PSpice, generate the output waveforms, and make any necessary corrections on the graph of Figure 8.6. Place an arrow at the exact time the parallel data (1101) is loaded.

Activity *JOHNSON*

Activity *JOHNSON* investigates the shift counter of Figure 8.7. By inverting and feeding the output back to the input, it assures continuous (counter) operation.

11. Add schematic *JOHNSON* to project *shift*.

12. Draw the circuit of Figure 8.7.

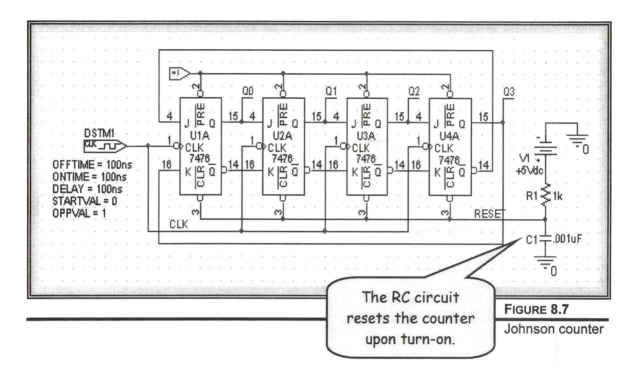

The RC circuit resets the counter upon turn-on.

FIGURE 8.7

Johnson counter

13. On the graph of Figure 8.8, predict (draw) the output waveforms (*Q0 to Q3*).

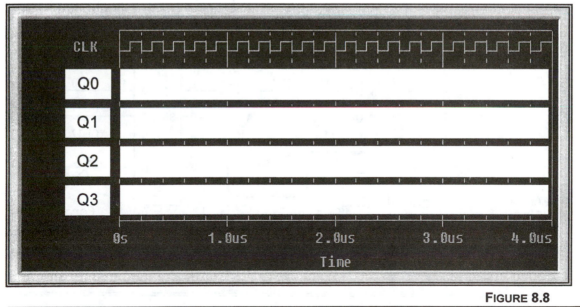

FIGURE 8.8

Johnson waveforms

14. Set the simulation profile to *Transient*, from 0 to 4μs, step ceiling 4ns, *typical* timing mode, flip-flops initialized to zero.

15. Run PSpice and generate the output waveforms. Make any necessary corrections on the graph of Figure 8.8.

Advanced Activities

Figure 8.9 is a complete serial communications system. Parallel data is loaded into U1 and walked out *QA* in serial. The serial data is input to U2 via the *SL* (shift-left) pin, shifted to the left, and output in parallel.

16. Create an appropriate schematic and draw the communication circuit of Figure 8.9. If using voltage markers (as shown), be sure to set them in the following order: CLOCK, IN[3-0], LOAD, SERIAL, and OUT[3-0].

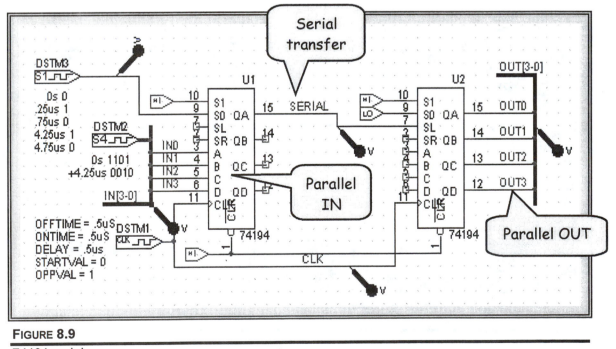

FIGURE 8.9

74194 serial communication test system

17. Set the simulation profile to *Transient* from 0 to 10µs, step ceiling 10ns, *typical* timing mode, flip-flops to zero.

18. Run PSpice and generate the waveforms of Figure 8.10. (Note that placing markers on bus strips causes waveforms to be displayed in hexadecimal format.)

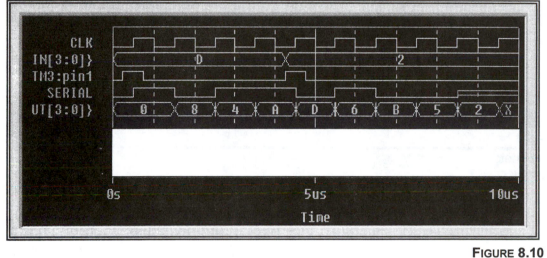

FIGURE 8.10
74194 system waveforms

19. Using the white space of Figure 8.10, add the following items:

   a. An arrow at the point where the first byte of input parallel data (1101) is loaded.

   b. A circle around the same data in serial form.

   c. An arrow at the point where the first byte of parallel data (1101) is output.

20. Reviewing Figure 8.10, explain why the output parallel data follows the pattern listed (0, 8, 4, A, D, 6, B, 5, and 2).

PSpice for Windows

Exercises

1. To rotate a stepper motor, we require the 4-bit sequence of Table 8.1. Show how to use a 74194 universal shift register to generate both sequences.

Clockwise Rotation	Counterclockwise Rotation
1000	0001
0100	0010
0010	0100
0001	1000

TABLE 8.1

Stepper motor logic sequences

2. Cascade two 74194 universal shift registers to perform 8-bit parallel to serial conversion. (*Hint*: Make use of the *SD* and *SL* serial data inputs.)

3. Design and test an 8-bit serial-to-parallel converter using the 74164 8-bit universal shift register.

Questions and Problems

1. Referring to the shift register of Figure 8.1, what is the state of the register after eight clock pulses?

2.   Is there such a thing as a *ripple* shift register?   (Are all shift registers synchronous?)

3.   Referring to the discrete P2S circuit of Figure 8.3, what would the output waveform (at *Q0*) look like if the JK settings to the initial flip-flop (*U1A*) were reversed (*J* HI and *K* LO)?

4.   How is the shift register of Figure 8.3 cleared (all outputs driven to zero)?

5.   Referring to the integrated shift register of Figure 8.5:

   a. What is the purpose of pins *SL* and *SR*?

   b. What happens to the output when S0 and S1 are held low?

6. When a logic HI is shifted out the right side of the Johnson counter of Figure 8.7, what is automatically shifted in the left side?

7. Referring to Figure 8.9, what is the advantage of transferring data in series?

8. What changes would you make in the communication circuit of Figure 8.9 to reverse the flow of data?

9

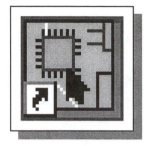

Digital Coding

Comparators, Coders, and Multiplexers

Objectives

- *To design and test circuits that compare binary inputs*
- *To design and test circuits that convert between decimal and BCD*
- *To design and test circuits that switch between multiple serial lines*

Discussion

In this chapter we investigate three popular types of digital discrete and integrated circuits:

- *Comparators*, which input two digital numbers and report if the first is greater than, equal to, or less than the second.

- *Decoders*, which convert a multibit input code into a single output line, and *encoders*, which reverse the process.

- *Multiplexers*, which select one signal from a number of choices, and *demultiplexers*, which reverse the process.

Simulation practice

Activity *COMPARATOR*

- Activity *COMPD* (comparator discrete) is based on the discrete 2-bit comparator of Figure 9.1.

- Schematic *COMPI* (comparator integrated) shows the same process accomplished with the integrated circuit (7485) of Figure 9.2.

1. Create project *ccm* with schematic *COMPD*.

2. Draw the discrete 2-bit comparator test circuit of Figure 9.1.

3. On the plot of Figure 9.3, draw the predicted output signal. (*Hint*: The inputs match when $D_0 = D_1$ and $D_2 = D_3$.)

4. Set the simulation profile to *Transient*, from 0 to 1µs, step-ceiling 1ns, *typical* timing mode, all flip-flops initialized to zero.

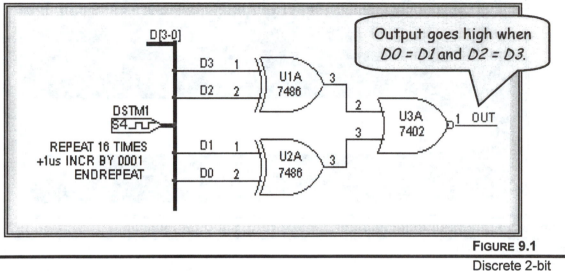

FIGURE 9.1

Discrete 2-bit comparator

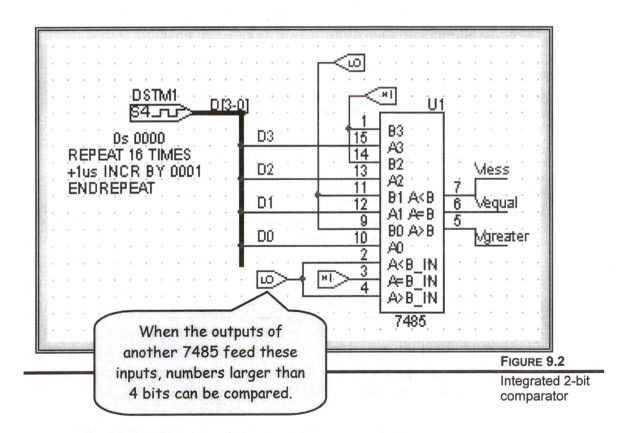

FIGURE 9.2

Integrated 2-bit comparator

PSpice for Windows

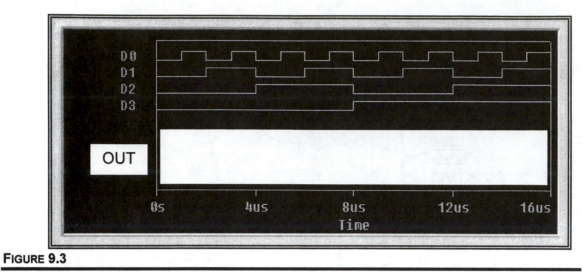

FIGURE 9.3

Discrete comparator
waveforms

5. Generate the waveforms using PSpice and compare them to your predictions. Were they the same? (Make any necessary corrections.)

Yes   No

Schematic *COMPI*

6. Add schematic *COMPI* to project *ccm*.

7. Draw the circuit of Figure 9.2.

> The 7485 compares the 4-bit values of the A and B inputs and activates the appropriate output line (A<B, A=B, A>B). By using the A<B_IN, A>B_IN, and A=B lines, any number of 7485s can be cascaded.

8. Given the input waveforms of Figure 9.4 (and the B inputs set as shown), predict (draw) the output waveforms (*Vless*, *Vequal*, and *Vgreater*) in the spaces provided.

9. Set the simulation profile for *Transient*, from 0 to 16µs, step ceiling 16ns, *typical* timing mode, and all flip-flops to zero.

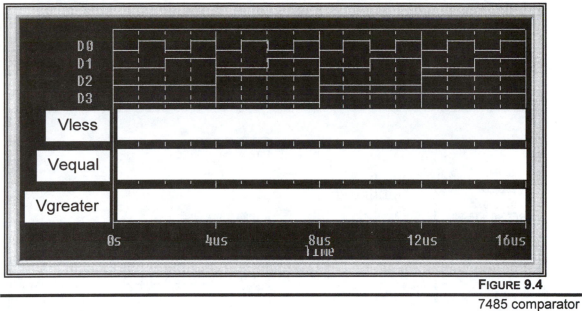

FIGURE 9.4

7485 comparator
waveforms

10. Generate the waveforms using PSpice and compare them to
     your predictions. Are they the same? (Make any necessary
     corrections.)

         Yes   No

Activity *DECODER/ENCODER*

This activity follows the previous discrete/integrated comparison
pattern:

- Activity *DED* (decoder/encoder *discrete*) combines the discrete
  encoder and decoder in the single schematic of Figure 9.5.

- Activity *DEI* (decoder/encoder *integrated*) combines the
  integrated encoder and decoder in the schematic of Figure 9.6.

Discrete Decoder/Encoder

11. Add schematic *DED* to project *ccm*.

12. Draw the discrete 2-bit decoder/encoder combination circuit of
     Figure 9.5.

In both cases, the ZERO output line of the decoder has no connection because the encoder output is automatically zero when all the remaining decoder outputs are zero.

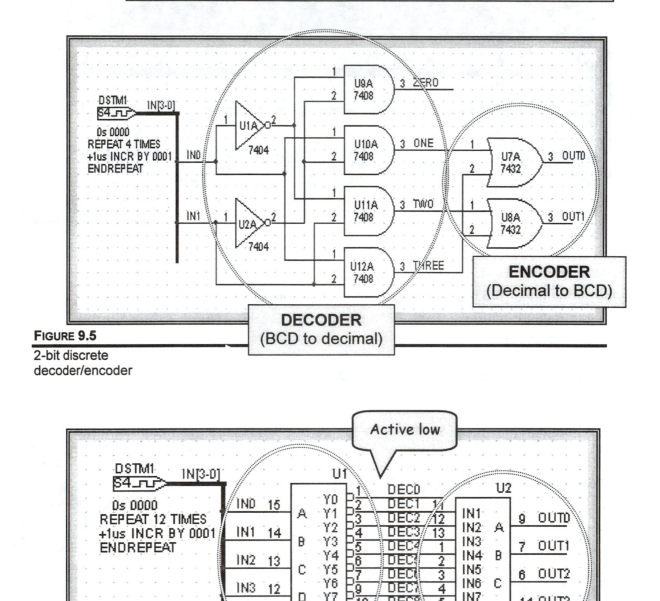

FIGURE 9.5

2-bit discrete
decoder/encoder

FIGURE 9.6

7442A/74147
integrated
decoder/encoder

PSpice for Windows

13. Realizing that the inputs and outputs are active high, predict and sketch the circuit outputs on the graph of Figure 9.7.

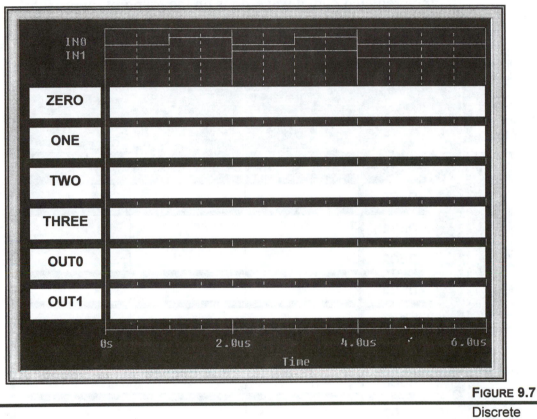

FIGURE 9.7

Discrete decoder/encoder waveforms

14. Set the simulation profile to *Transient*, from 0 to 6μs, step ceiling 6ns, *typical* timing mode, all flip-flops to zero.

15. Run PSpice and generate the set of waveforms. Were your predictions correct? (Make any necessary changes.)

    Yes   No

16. Did the 0,0 input correctly decode/encode back to 0,0, despite the fact that line ZERO is left floating?

    Yes   No

Integrated Decoder/Encoder

17. Add schematic *DEI* (decoder/encoder integrated) to project *ccm*, draw the decoder/encoder circuit of Figure 9.6, and sketch the predicted output waveforms on the graph of Figure 9.8.

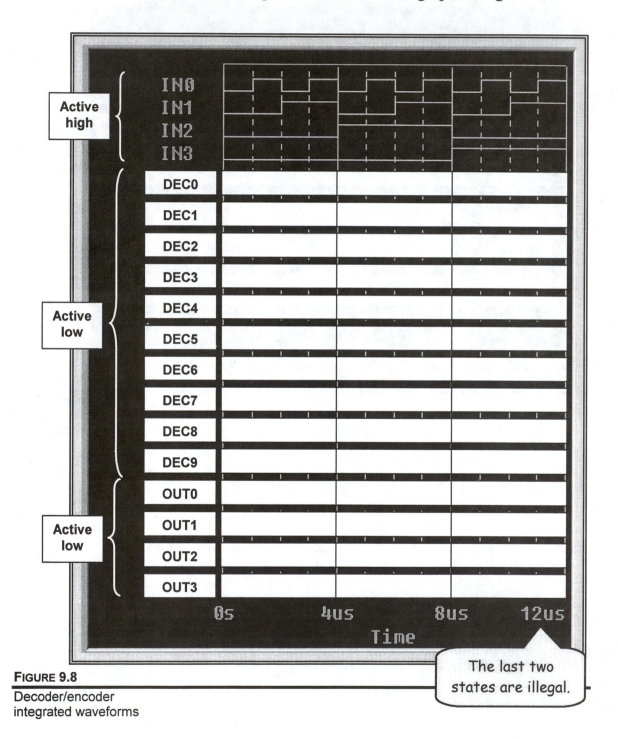

FIGURE 9.8

Decoder/encoder integrated waveforms

PSpice for Windows

18. Set the simulation profile for *Transient*, from 0 to 4μs, step ceiling 4ns, *typical* timing mode, all flip-flops to zero.

19. Run PSpice and generate the waveform set of Figure 9.8.

    a. Were your predictions correct? (Correct any mistakes.)

        Yes   No

    b. Was the output (*OUT0 – OUT3*) inverted from the input (*IN0 – IN3*)?

        Yes   No

    c. Were the decoder outputs (*DEC0 – DEC9*) active low?

        Yes   No

    d. Were the encoder outputs high for all "illegal" inputs (above 9)?

        Yes   No

Activity *MUX/DEMUX*

Again, following the previous pattern:

- Activity *MDD* (multiplex/demultiplex discrete) is based on the discrete circuit of Figure 9.9.

- Activity *MDI* (multiplex/demultiplex integrated) is based on the integrated circuit of Figure 9.10.

Discrete Multiplexer/Demultiplexer

20. Add schematic *MDD* to project *ccm*.

21. Draw the 2-bit discrete mux/demux circuit of Figure 9.9.

22. Given the circuit inputs of Figure 9.11, predict (draw) the expected waveforms listed.

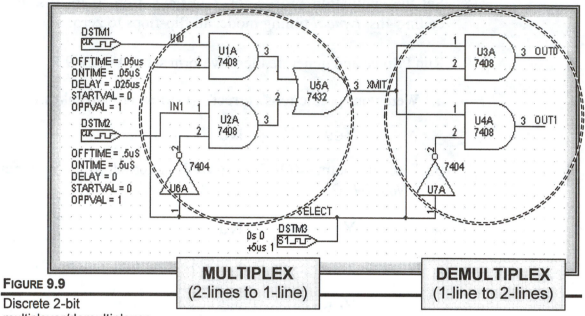

FIGURE 9.9

Discrete 2-bit
multiplexer/demultiplexer

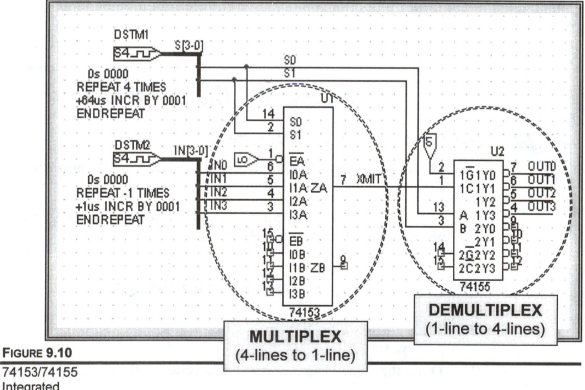

FIGURE 9.10

74153/74155
Integrated
mux/demux

PSpice for Windows

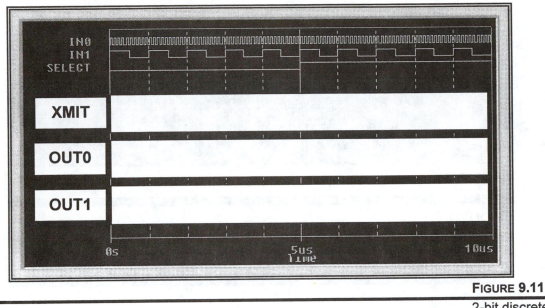

FIGURE 9.11

2-bit discrete
mux/demux
waveforms

23. Set the simulation profile for *Transient*, from 0 to 10μs, step ceiling 10ns, *typical* timing, flip-flops to zero.

24. Run PSpice and generate the waveforms of Figure 9.11. Were your predictions correct? (Make any necessary corrections.)

    Yes   No

Integrated Multiplexer/Demultiplexer

25. Add schematic *MDI* to project *ccm*, and draw the decoder/encoder circuit of Figure 9.10.

26. Given the circuit inputs on the graph of Figure 9.12, predict (draw) the expected waveforms listed.

27. Set the simulation profile for *Transient*, from 0 to 300μs, run PSpice, and generate the waveform set of Figure 9.12. Were your predictions correct? (Make any necessary changes.)

    Yes   No

PSpice for Windows

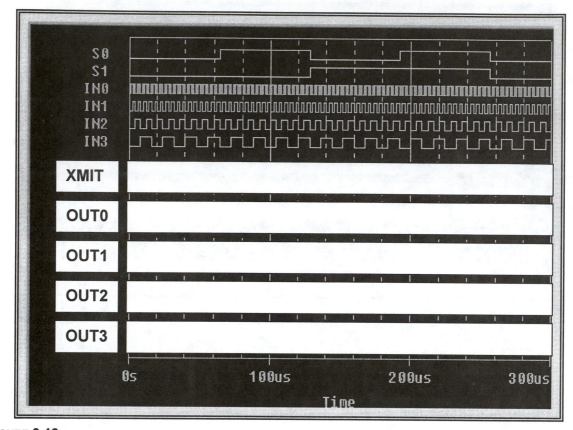

FIGURE 9.12

4-bit integrated
74153/74155
mux/demux waveforms

Advanced Activities

28. Add a "greater-than" output to the simple 2-bit comparator of Figure 9.1.

29. Modify the 2-bit decoder circuit of Figure 9.5 so both inputs and outputs are active low.

30. Upgrade the 2-bit multiplexer/demultiplexer of Figure 9.9 to a 3-bit system.

31. Show how to perform parallel-to-serial conversion using the 74153 multiplexer of Figure 9.10.

Exercises

1. Using the appropriate circuits of this chapter, design an alarm system for a room with eight doors and windows. Show how the system might interface to a microprocessor.

2. Using the techniques of this chapter, design a circuit that will go HIGH whenever the 4-bit binary input is greater than 7.

3. Using the integrated circuits of this chapter (and previous chapters as needed), design a UART (universal asynchronous receiver-transmitter). Your design will input a 4-bit data word in parallel and transmit the word in serial with the active-low start bit in the front and two active-high stop bits at the end.

Questions and Problems

1. When output *Vless* of Figure 9.2 goes active high, what do we know about the *A* and *B* inputs?

2. What is the difference between an *encoder* and a *decoder*?

3. When we wish to change a 10-line decimal input to BCD, we look for an:

    Encoder   Decoder

4. Why is there no connection from the ZERO line of the decoder to the encoder of Figures 9.5 and 9.6?

5. Why is the 74147 called a *priority* encoder?   (*Hint*: Examine the 74147's data sheet in Appendix C.)

6. When using the 74153 multiplexer of Figure 9.10, can two completely independent multiplex operations occur simultaneously?

7.   Referring to Figure 9.6, what happens to the outputs of the 7442A decoder and
     74147 encoder when the input is above 9?

8.   In the *PSTN* (public-switched telephone network), all phones are sampled at
     the rate of 8kHz.   How would you change the DSTM1 clock of Figure 9.10 to
     match this speed?

10

555 Timer

Multivibrator Operation

Objectives

- *To configure the 555 timer for continuous or one-time waveform generation*
- *To configure the 555 timer as a voltage-controlled oscillator*

Discussion

The 555 timer is another popular analog/digital integrated circuit (like the op amp). This versatile chip is used for a wide variety of oscillation and timing needs. The internal schematic of the 555 is shown on Figure 10.1. Note that the 555 uses a combination of analog and digital parts.

By adding external components, the 555 can be configured as any of the following:

- Astable multivibrator
- Monostable multivibrator
- Voltage-controlled oscillator

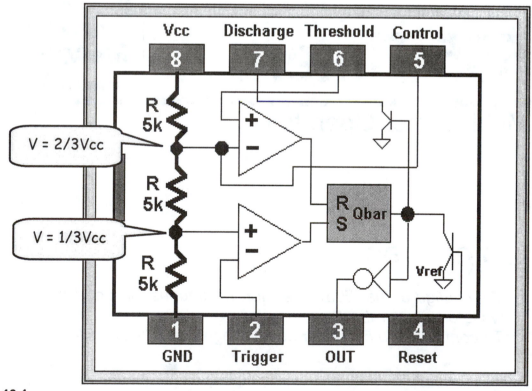

FIGURE 10.1

555 timer
schematics

Simulation Practice

Activity *MULTIVIBRATOR*

Activity MULTIVIBRATOR sets up the 555 timer of Figure 10.2 for continuous square wave (astable) operation.

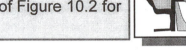

1.   Create project *555* with schematic *ASTABLE*.

2.   Draw the circuit of Figure 10.2.

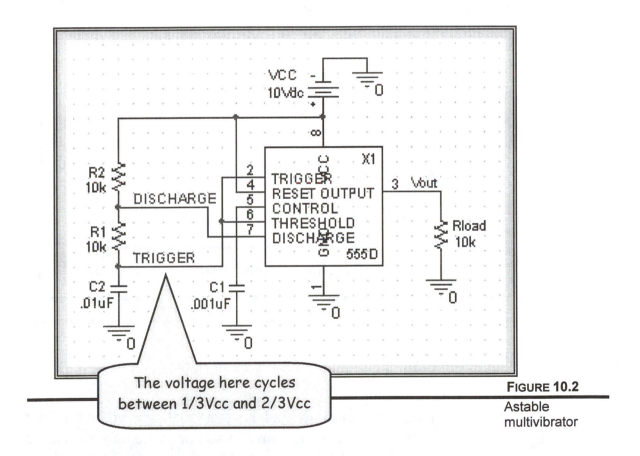

FIGURE 10.2

Astable multivibrator

3.   Set the simulation profile for *Transient* from 0 to .6ms, step ceiling of .6us, typical timing, and internal flip-flop initialized to one.

4. Using the equations below, calculate the expected frequency and duty cycle. Using the results, draw the predicted trigger (*VT*) and output (*V_OUT*) waveforms on the graph of Figure 10.3. (Be aware that the *TRIGGER* voltage starts at 10V, and assumes a regular pattern after the first cycle.)

$$f_0 \text{(frequency)} = \frac{1.44}{(R_A + 2R_B)C1} = \underline{\hspace{2cm}}$$

$$DC \text{ (\% duty cycle)} = \frac{(R_A + R_B)}{(R_A + 2R_B)} \times 100 = \underline{\hspace{2cm}}$$

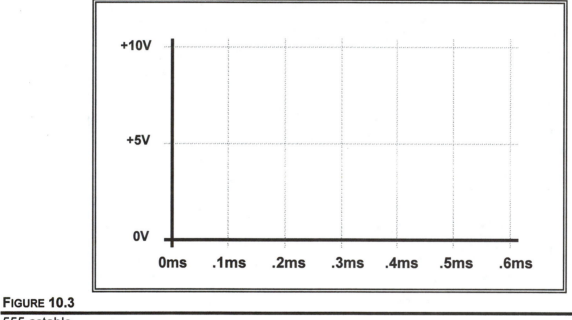

FIGURE 10.3

555 astable waveforms

5. Run a transient analysis and generate the trigger and output waveforms. Do the actual waveforms approximately match the expected values? Make any necessary changes to Figure 10.3.

        Yes   No

Monostable Multivibrator

6. By freeing up the trigger (pin 2), configure the system for monostable operation (Figure 10.4).

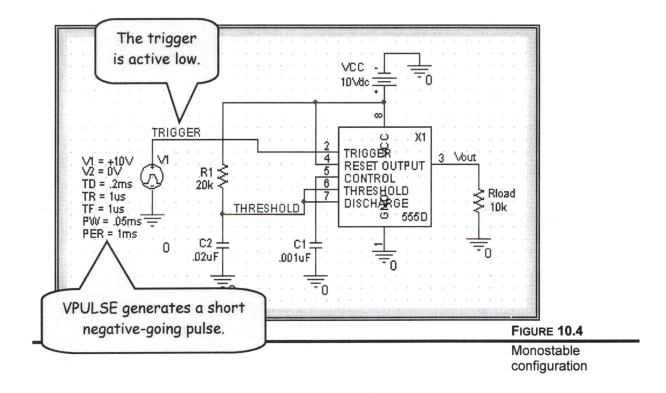

FIGURE 10.4

Monostable configuration

7. When the trigger goes low, the output goes high for a period of time determined by the equation PW = 1.1RC. Using this equation, draw the expected trigger (*TRIGGER*), threshold (*THRESHOLD*), and output (*Vout*) waveforms on the graph of Figure 10.5.

8. Set the simulation profile to *Transient* from 0 to 2ms, step ceiling 2us, typical timing, flip-flop initialized to one.

9. Run PSpice and display the *trigger*, *threshold*, and *output* waveforms on separate plots. Do the actual waveforms approximately match the expected values? (Make any necessary corrections.)

Yes   No

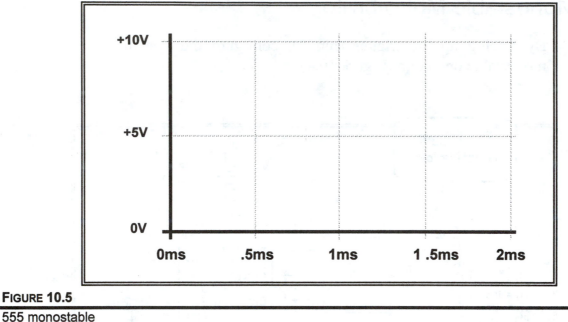

FIGURE 10.5

555 monostable
waveforms

Advanced Activities

10. By varying the voltage into pin 5, we create a voltage-controlled oscillator (VCO).   Set up the 555 for VCO operation as shown in Figure 10.6.

11. Modify the simulation profile for transient operation from 0 to 3ms.

12. Run PSpice and generate the control (*CONTROL*), trigger (*TRIGGER*), and output (Vout) waveforms of Figure 10.7.

13. Based on the results, what output frequency corresponds *approximately* to each of the following control voltages?

   a. f(out) at 2V ≅ _____

   b. f(out) at 6V ≅ _____

   c. f(out) at 10V ≅ _____

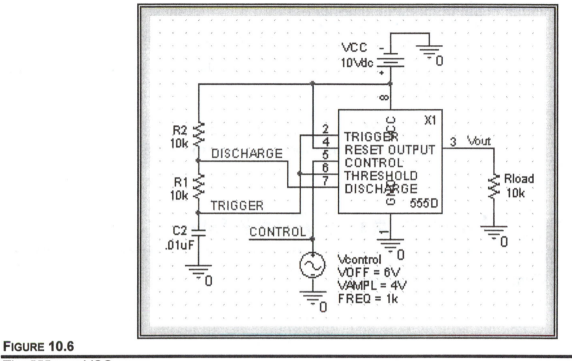

FIGURE 10.6

The 555 as a VCO

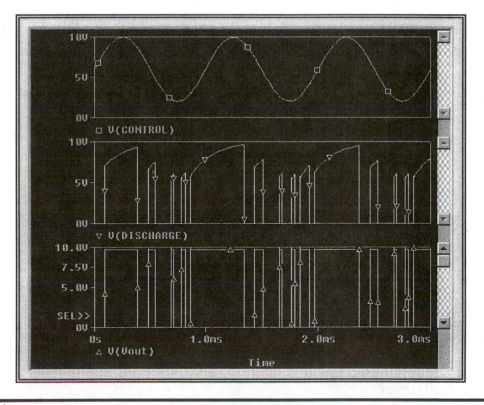

FIGURE 10.7

VCO control and
output waveforms

PSpice for Windows

Exercises

1. Using the 555, design a circuit to delay exactly 1 second from the time of a trigger.

2. Using the 555, design a circuit to produce a 1kHz waveform of 25% duty cycle (high for .25ms, low for .75ms).

Questions and Problems

1. What does *monostable* mean?

2. Refer to Figure 10.1. What trigger and threshold voltages (as a percentage of Vcc) will set and reset the flip-flop?

3. Is the trigger signal of the monostable multivibrator of Figure 10.4 active high or active low?

4. Which of the following 555 signals reacts to a *rising* high-level voltage?
    a. Threshold
    b. Trigger

5. Refer to Figure 10.1.   How does the control voltage (pin 5) affect the frequency of operation?

6. In Chapter 14, we will use a 555 timer to create a PLL. What does *PLL* stand for?

Part 4

Converters and RAM

The real world in which we all live is primarily an analog world. A computer, on the other hand, works only with the ones and zeros of its purely digital environment. For our computer to be a truly useful servant, it must listen to our commands, it must remember our wishes, and it must respond accordingly.

In the two chapters of Part 4, we learn to translate between analog and digital and we learn how information is stored in digital memory.

11

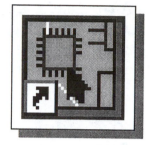

Analog/ Digital Conversions

Resolution

Objectives

- *To convert analog signals to digital (ADC)*
- *To convert digital signals to analog (DAC)*
- *To compare discrete and integrated ADCs and DACs*

Discussion

A computer is a digital "alien" in an analog world. To carry out our wishes, the computer must intake analog information (*analog-to-digital*), process it in digital form, and return the result to us in analog form (*digital-to-analog*).

The simplest of the two operations is digital-to-analog conversion (*DAC*). (It is easier for a computer to talk to us than vice versa.) The two DAC methods covered by this chapter are:

- Binary weighted
- Use of integrated circuit

The more difficult analog-to-digital conversion (*ADC*) is accomplished in several ways. The two ways covered by this chapter are:

- Counter-ramp (staircase)
- Use of integrated circuit

As we will see, the use of an integrated circuit greatly improves convenience and precision.

Simulation Practice

Activity *DIGITAL2ANALOG*

Following the lead of Chapter 9, we present both discrete and integrated versions:

- Schematic *DACD* (digital-to-analog discrete) uses the binary-weighted process of Figure 11.1 to convert a 4-bit digital word to analog.

- Schematic *DACI* (digital-to-analog integrated) uses the integrated circuit of Figure 11.2 to convert an 8-bit word to analog.

Discrete DAC

1. Create project *analogdigital* with schematic *DACD* (digital-to-analog discrete), and draw the circuit of Figure 11.1.

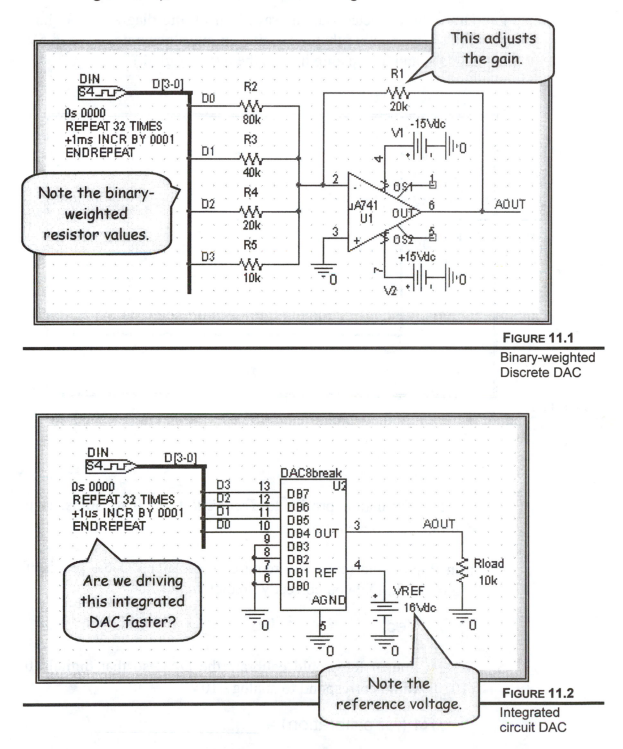

FIGURE 11.1

Binary-weighted
Discrete DAC

FIGURE 11.2

Integrated
circuit DAC

PSpice for Windows

When using this discrete DAC technique, digitally-generated currents are weighted in the 8, 4, 2, 1 digital pattern, summed together, and converted to an output analog voltage.

2. Draw the expected output waveform on the diagram of Figure 11.3.   (Remember that the digital stimulus generates 0V and 5V signals, and the op amp inverts the output signal.)

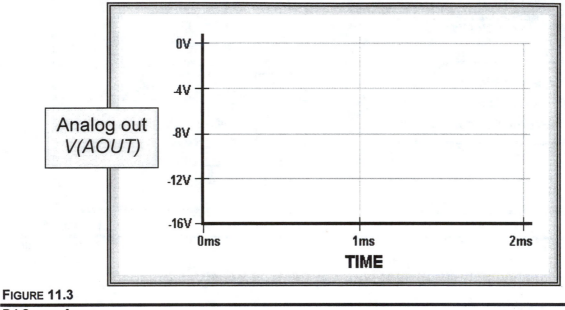

FIGURE 11.3

DAC waveforms

3. Set the simulation profile to *Transient* from 0 to 30ms, step ceiling 30µs, *typical* timing.

4. Run PSpice and make any required corrections to the graph of Figure 11.3. Did you correctly predict the results?

    Yes   No

5. What value of R1 would calibrate the DAC so that digital 10 (1010) would correspond to analog −10V?

    R1 (for calibration) = _____

6. Substitute the value of R1 determined in step 4 and run a simulation. Is the DAC now calibrated properly? (Be aware that the −14.7V rail may be hit before −15V is reached.)

      Yes   No

Integrated Circuit DAC

7. Add schematic *DACI* (digital-to-analog integrated) to project *analogdigital*.

8. Draw the integrated circuit DAC system of Figure 11.2. (Part *DAC8break* is found in the *BREAKOUT* library).

9. Based on the equation below (and noting the reference voltage on Figure 11.2), predict the output of the DAC and draw your results on the graph of Figure 11.4.

$$V_{AOUT} = V(ref, gnd) \times \frac{(binary\ value\ of\ inputs)}{2^8}$$

For example, 1,0,0,0 = 128

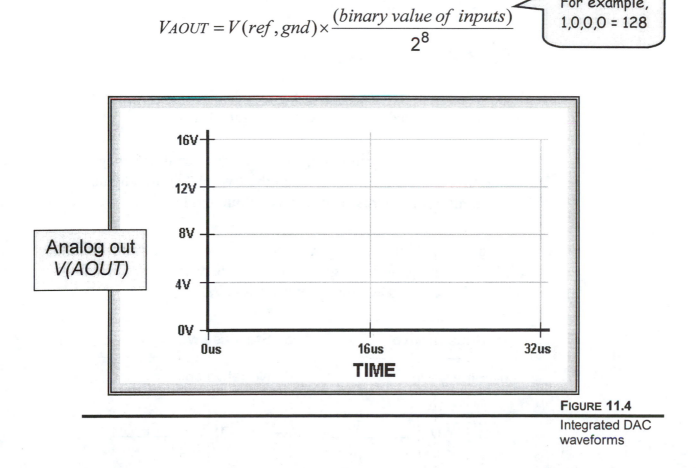

FIGURE 11.4

Integrated DAC waveforms

10. Set the simulation profile for *Transient* from 0 to 40μs, step ceiling 40ns, typical timing, flip-flops initialized to zero.

11. Run PSpice and draw the actual DAC output waveform. Did it match your predictions?

Yes   No

Activity *ANALOG2DIGITAL*

Once again, we present both discrete and integrated versions:

- Schematic *ADCD* (analog-to-digital conversion discrete) uses the staircase method of Figure 11.5 to convert an analog input to a 4-bit digital output.

- Schematic *ADCI* (analog-to-digital conversion integrated) uses the single integrated circuit of Figure 11.6 to convert an 8-bit digital word to analog.

Analog-to-Digital Discrete

12. Add schematic *ADCD* (analog-to-digital discrete) to project *analogdigital*, and draw the circuit of Figure 11.5.

> Our discrete ADC circuit compares the input analog voltage to a rising staircase voltage. When the two match, the digital equivalent of the staircase voltage is output and latched.

13. Instead of sketching all the ADC's waveforms (a formidable task), answer the following: Which of the waveforms listed below will be found at *V(DAC)* and *V(LATCH)*?

a. Square wave	c. Staircase wave
b. Sine wave	d. Negative spice

V(DAC) = _____   V(LATCH) = _____

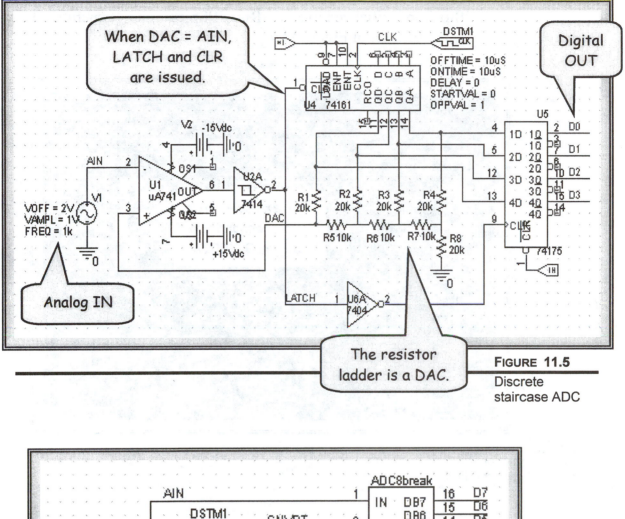

FIGURE 11.5

Discrete staircase ADC

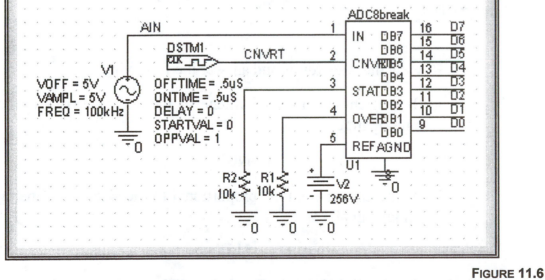

FIGURE 11.6

Integrated circuit ADC

14. Set the simulation profile for *Transient* from 0 to 2ms, with all flip-flops initialized to zero. Run PSpice and generate the waveform set of Figure 11.7.

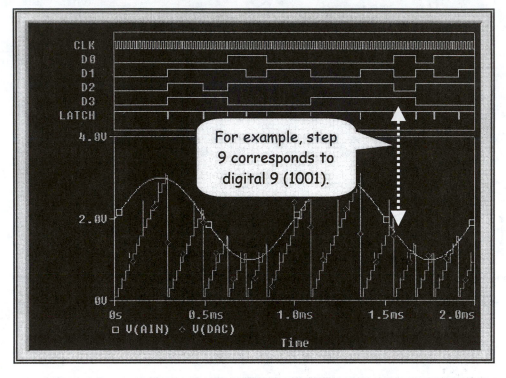

FIGURE 11.7

ADC waveforms for discrete staircase method

15. Review the waveforms and answer the following:

a. Is a latch signal issued whenever the staircase rises to the analog input level?

        Yes   No

b. Is the conversion time proportional to the size of the analog input signal?

        Yes   No

c. Do the digital output signals match the analog input?

        Yes   No

16. To test the resolution of the discrete ADC, add a DAC to the output and generate the waveforms of Figure 11.8. (*Hint*: Because part *DAC8break* exceeds the evaluation limit, use the resistor ladder shown in Figure 11.5 to perform DAC services.)

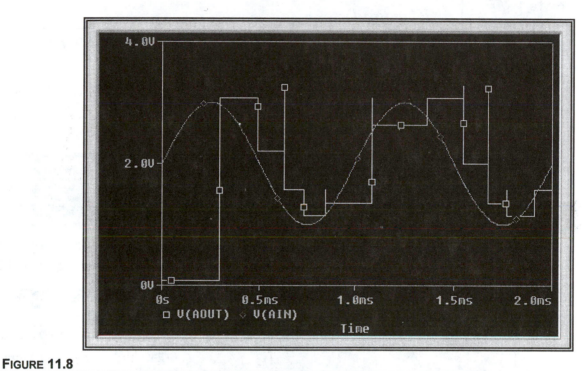

FIGURE 11.8

A/D resolution

17. Is the resolution poor?

    Yes   No

18. To improve the resolution, double the clock speed of DSTM1. Was the resolution improved?

    Yes   No

Integrated Circuit ADC

19. Add schematic *ADCI* (analog-to-digital integrated) to project *analogdigital*.

20. Draw the integrated circuit ADC system of Figure 11.6. (Part *ADC8break* is from the *BREAKOUT* library).

21. Based on the equation below, predict the digital (binary) output of the ADC when *AIN* is 10V. (*Hint*: The system is calibrated.)

Note *VREF* (V2) on Figure 11.6.

$$\frac{V(in, gnd)}{V(ref, gnd)} \times 2^8 = \frac{}{D7}\frac{}{D6}\frac{}{D5}\frac{}{D4}\frac{}{D3}\frac{}{D2}\frac{}{D1}\frac{}{D0}$$

22. Set the simulation profile, run PSpice, and generate the curves of Figure 11.9.

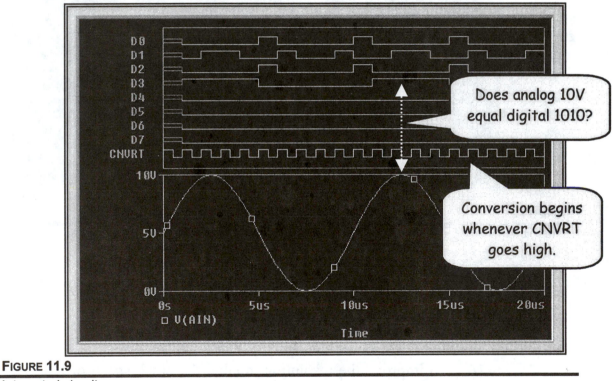

FIGURE 11.9

Integrated circuit
A/D waveforms

23. Refering to Figure 11.9, answer the following:

a. The integrated circuit ADC of Figure 11.6 ran faster than the discrete version of Figure 11.5.

       Yes   No

b. Do the digital outputs correspond to the analog input values? (Does 10V in equal 00001010 out?)

Yes   No

24. As before, test the resolution of our A/D converter by adding the DAC of Figure 11.10 to the output. (This time we can add the integrated circuit version without exceeding the evaluation limit.)

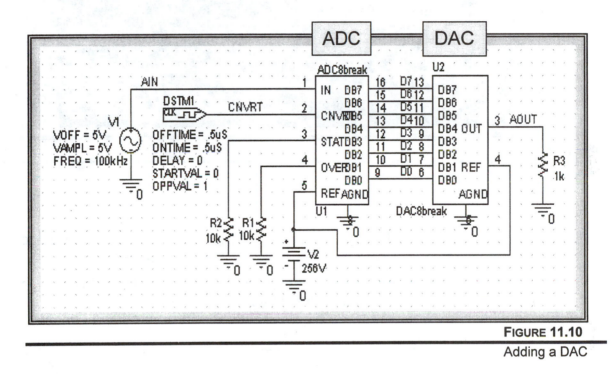

FIGURE 11.10

Adding a DAC

25. Run PSpice and generate the curves of Figure 11.11. Does the output equal the input, but is the resolution still quite low?

Yes   No

> As before, we could increase resolution by increasing clock speed—and we will. However, regardless of the clock speed, we would still be limited to 16 voltage levels. Therefore, to maximize resolution, we must use all 256 steps provided by this 8-bit device.

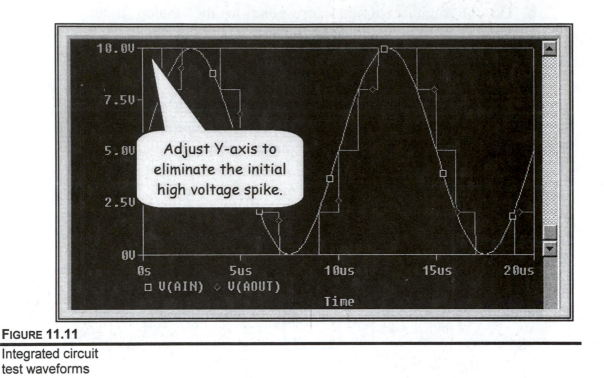

FIGURE 11.11

Integrated circuit
test waveforms

26. Based on the equation of step 21, what reference voltage value will make the maximum input analog voltage (10V) equivalent to the maximum binary output (11111111)? (*Hint*: 11111111_2 = 255_{10}, giving an answer slightly above 10V.)

V_{REF} = _____

27. Change V_{REF} to the value determined in step 26, increase the clock (*DSTM1*) frequency to 10MEGHz (ON and OFF times = .05μs) and generate the new waveform set of Figure 11.12.

28. Zoom in on the AIN/AOUT waveforms of Figure 11.12.

   a. Is *AOUT* a staircase of very small steps (.1 to .3V)?

   Yes   No

   b. Is analog 10V equal to digital 11111111?

   Yes   No

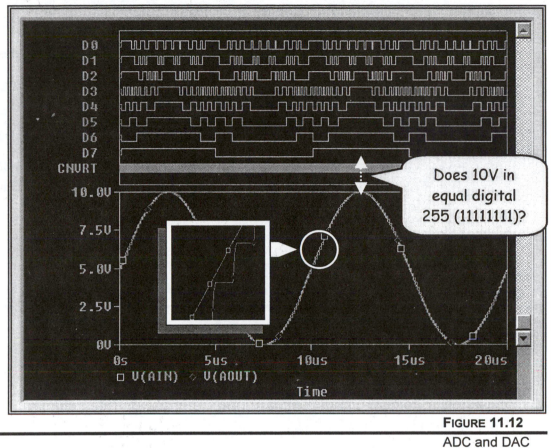

FIGURE 11.12
ADC and DAC
precision waveforms

Advanced Activities

29. Redesign the DAC circuit of Figure 11.1 using a resistor ladder input to a VCVS buffer op amp. (*Hint*: A DAC resistor ladder is shown in Figure 11.5.)

30. By double buffering the ADC ramp circuit of Figure 11.5 (using two latches), redesign the circuit so the conversion time is constant. (*Hint*: Let the conversion time be the time required for the counter to cycle fully from 0 to 15.)

31. Instead of analog-to-digital-to-analog (as shown in Figure 11.10), reverse the process and test the system.

Exercises

1. Using basic circuit elements (such as those of Figure 11.5), design an A/D converter based on the successive-approximation technique. (Research will be required.)

2. By experimentation, determine the highest speeds of operation for both the integrated-circuit DAC of Figure 11.2 and the integrated-circuit ADC of Figure 11.6.

Questions and Problems

1.   The operational amplifier of Figure 11.1 is a _____ -to- _____ converter.

2.   Based on the equation of step 9 for the integrated circuit D/A converter of Figure 11.2, what would the output voltage be if the binary input were 00001111?

3.   Regarding the counter-ramp (staircase) A/D converter of Figure 11.5, why do the conversion times vary?

4. Regarding the integrated circuit ADC converter of Figure 11.6, why is the reference voltage initially set at 256V?

5. What is the maximum (best) resolution of the 8-bit ADC of Figure 11.6?

6. A data acquisition system would sample analog voltages, convert them to digital, and store them for future reference.   What circuit component would accomplish the storage feature? (*Hint*: See Chapter 12.)

12

Random-Access Memory

Data Acquisition

Objectives

- *To write data to and read data from a random-access memory (RAM) chip*
- *To design a data acquisition system*
- *To display parallel data in hexadecimal*

Discussion

Random-access memory (RAM) is a *read/write volatile* storage device for digital data. Positive-feedback latching action maintains the digital states in internal latches—but all data is lost when power is removed. By following the proper timing requirements, data is *written* to and *read* from *addressable* locations.

A typical RAM chip is shown in Figure 12.1. It is an 8k × 8 device, meaning that it offers 8k (8,192) addressable memory locations, each storing 8 bits of data.

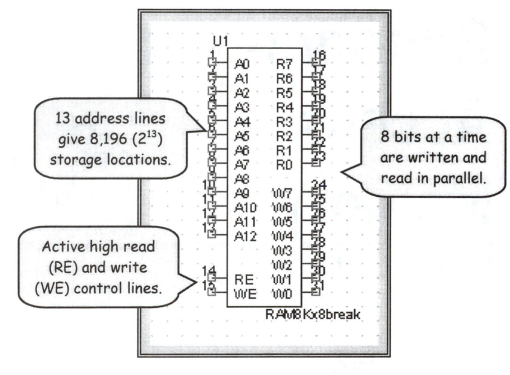

FIGURE 12.1

8k × 8 RAM device

Figure 12.2 shows the device timing specifications for the read and write operations. Note the narrow *WE* pulse during write operations. To avoid incorrect data going to the wrong memory location, it is especially critical that the address and data be stable when WE is active high.

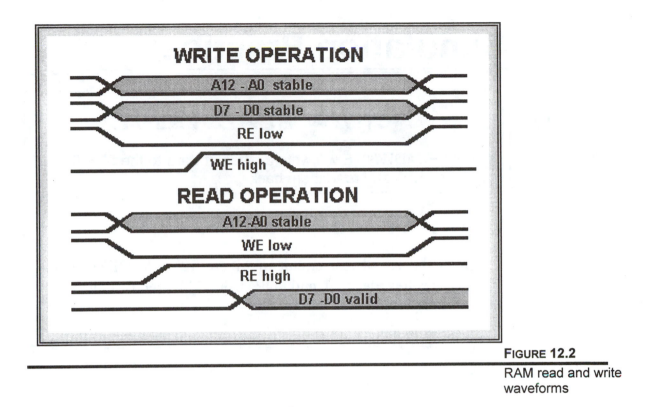

FIGURE 12.2

RAM read and write waveforms

Data Acquisition

A *data acquisition system* samples and stores analog values for later playback. The most widely used storage medium is *random-access memory* (RAM). In a typical case, a data acquisition system might be used to store the winter temperature readings at a remote arctic site. When spring comes, the data can be retrieved for analysis.

Data acquisition and retrieval is performed as follows, where *n* is the number of RAM storage locations used:

Repeat *n* times to store:
1. Sample an analog signal.
2. Convert to digital.
3. Write to the next RAM location.

Repeat *n* times to retrieve:
1. Read from the next RAM location.
2. Convert to analog.
3. Display analog signal.

Simulation Practice

Activity *READWRITE*

Activity *READWRITE* will write 2 bytes of data to the 8k × 8 RAM of Figure 12.3 and read them back.

1. Create project *ram* with schematic *READWRITE*.

2. Draw the RAM test circuit of Figure 12.3. (FORMAT 4444 specifies hexadecimal, and FORMAT 1111 specifies binary.)

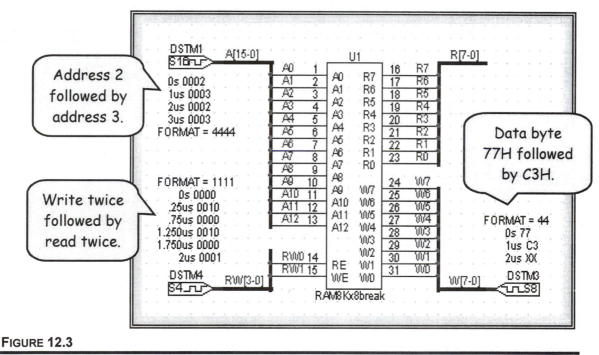

FIGURE 12.3

RAM test circuits

3. Set the simulation profile for *Transient* from 0 to 4µs, step ceiling 4ns, *typical* timing, initialize flip-flops to zero.

4. Run PSpice and generate the hexidecimal waveform set of Figure 12.4.

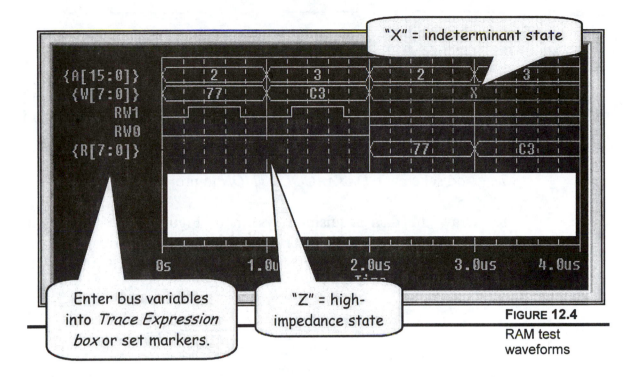

FIGURE **12.4**

RAM test waveforms

5. In the white space of Figure 12.4, use brackets and labels to show when the two write and two read processes take place.

6. Based on the results of Figure 12.4:

a. Do the read/write waveforms follow the standards of Figure 12.2?

Yes   No

b. Were data bytes 77 and C3 written to locations 2 and 3 and read back?

Yes   No

c. We could have displayed all waveforms in binary rather than hexadecimal. Why is hexadecimal more convenient?

PSpice for Windows

Activity *DATAACQUISITION*

Activity *DATAACQUISITION* will use the circuit of Figure 12.5 to sample and store analog signals from a remote sensor in RAM for future playback. Integrated circuit DAC and ADC devices interface the digital RAM to the analog world.

7. Add schematic *DATAACQUISITION* to project *ram*.

8. Draw the data acquisition system of Figure 12.5. Note that *VREF* (*V2*) is calibrated for direct analog/digital correspondence as introduced in Chapter 11. (That is, 10V in equals 00001010 out.)

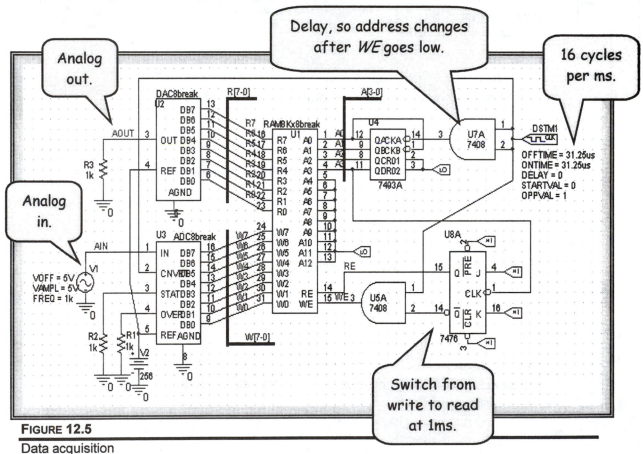

FIGURE 12.5

Data acquisition system

Briefly, the system works as follows:

- A sine wave analog signal enters the system at line AIN. Each time CLOCK goes high the waveform is sampled and converted to digital. After a short delay, WE (write) cycles high and low and the data word is written to the RAM. The address is incremented by the counter and the process repeats. After 16 WRITE cycles, the counter overflows, the flip-flip is toggled, and we enter the READ process.

- During each READ cycle, RE stays high and WE remains low. Data is output to the DAC through R7 to R0, converted to analog, and presented at output line AOUT. The process is repeated for 16 CLOCK cycles until the flip-flop again toggles and we return to the WRITE operation.

9. Run PSpice and generate the input/output analog/digital waveform set of Figure 12.6.

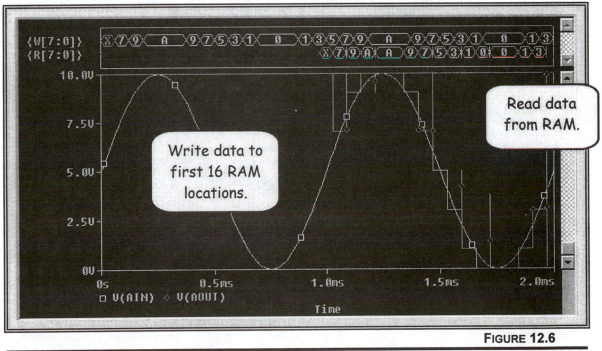

FIGURE 12.6

Analog/digital data acquisition waveforms

10. Reviewing the analog waveform set of Figure 12.6, did the data acquisition system sample the analog input 16 times and store the 16 digital equivalents properly? Did it then output each of the 16 stored words to the DAC and generate the proper waveform?

Yes   No

11. Making use of the individual net aliases, generate the digital (binary) waveform set of Figure 12.7.

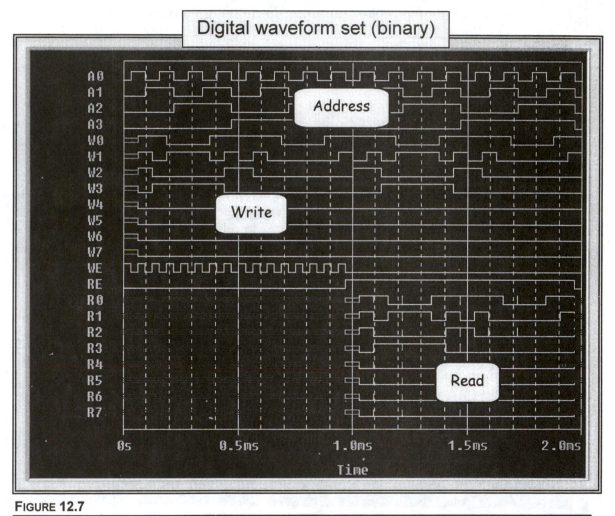

FIGURE 12.7

Binary format
digital data
acquisition waveforms

PSpice for Windows

12. Reviewing the waveforms of Figure 12.7:

    a. Did the address lines (A3 to A0) cycle from 0 to 15?

        Yes   No

    b. Are the READ waveforms identical to the WRITE waveforms?

        Yes   No

    c. Do the READ/WRITE waveforms match the specifications of Figure 12.2?

        Yes   No

13. This time, use the bus strip signals shown in Figure 12.5 to generate the hexadecimal format waveform set of Figure 12.8.

    a. The upper curve shows asterisks for most data states because there is insufficient space to write the numerical values.

    b. By expanding the waveforms, as shown by the lower curve, sufficient room is created and the values are displayed.

14. Are the hexadecimal format waveforms of Figure 12.8 equivalent to the binary format waveforms of Figure 12.7?

        Yes   No

Advanced Activities

15. Redesign the data acquisition system of Figure 12.5 to increase resolution by making use of the first 256 memory locations. (*Hint*: Change *VREF* to 10.04V, cascade two 7493A counters, and increase the clock speed.)

16. Test the *Nyquist theorem* on an input square wave. (To avoid losing information, the input analog signal must be sampled at a rate that is at least twice the highest input frequency of interest.)

PSpice for Windows

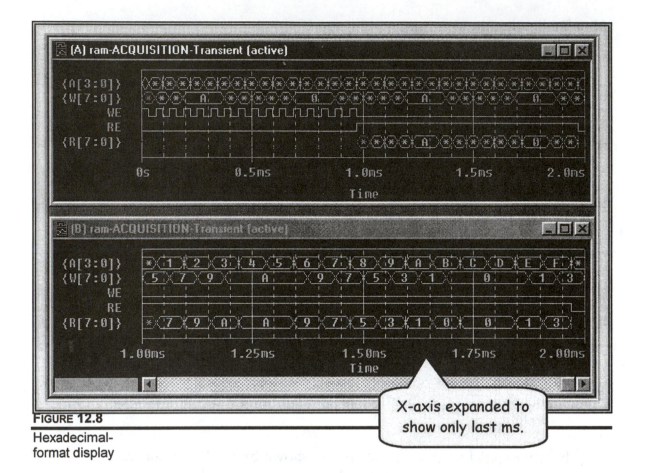

FIGURE 12.8

Hexadecimal-
format display

X-axis expanded to
show only last ms.

17. By modifying the data acquisition system of Figure 12.5, design an advanced system that uses time multiplexing to alternately sample, store, and play back *two* analog inputs.

18. Using part *Rbreak*, design a temperature transducer for use in a data acquisition system in which output voltage is proportional to temperature.

Exercises

1. Design a system that erases all 8,192 RAM locations.

2. By making use of a serial-to-parallel converter (Chapter 8), redesign the data acquisition system of Figure 12.5 for digital serial input.

Questions and Problems

1. How many bits of data can be stored in a 32k × 4 RAM?

2. During a write operation, why must *WE* be low when the address is changing?

3. Referring to the READ operation of Figure 12.2, why is there a lag between the time RE goes high and the data is valid?

4. Under what conditions would a data acquisition system not require a DAC or ADC?

5. How would the analog output waveform (AOUT) of Figure 12.6 change if VREF were doubled to 20.08V?

6.   During a WRITE operation, when RE is low, in what state are the read outputs (R7 to R0)?

Part 5

Data Communications

We communicate by speaking. It is a simple and direct process. However, what if a vast ocean separates the talker and the listener? The only practical way to communicate is by electronic wizardry.

In the two chapters of Part 5, we examine two diverse parts of the vast communication landscape: amplitude modulation and the phase-locked loop (PLL).

CHAPTER

13

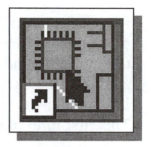

Amplitude
Modulation

Detection

Objectives

- *To generate an amplitude-modulated (AM) signal*
- *To detect an AM signal*
- *To display the frequency-domain waveforms of an AM system*

Discussion

An electromagnetic wave is generated whenever an electron is accelerated. However, transmission of electromagnetic waves through the atmosphere is efficient only at frequencies well above audio. This fact of life led to the concept of a low-frequency, information-carrying signal *modulating* a high-frequency *carrier*. This first form of modulation was *amplitude modulation*.

Amplitude Modulation

A signal is amplitude modulated when a low-frequency information (voice) signal controls the amplitude of a high-frequency carrier. A bipolar transistor can be used to modulate a signal because its gain depends on bias current ($A = rload/re'$ and $re' \cong 25mV/I_{EQ}$).

That is, the low-frequency information signal changes the bias current (I_{EQ}), which changes re', which changes the gain ($rload/re'$), which finally changes the amplitude of the transmitted carrier.

The term *percent modulation* is a measure of the strength of the modulating signal. It is defined as follows:

$$\% \, Modulation = \frac{Maximum \, gain - Minimum \, gain}{Nominal \, gain \, (no \, modulation \, signal)} \times 100$$

Linear Versus Nonlinear

When the modulation circuit is nonlinear, the resulting modulated signal contains *sidebands* equal to all multiples of the sum and difference of the carrier and modulation signals. A Fourier analysis of the output signal will reveal these information-carrying frequency components.

When the modulated signal is received, it must be *demodulated* (the information signal must be extracted from the carrier). The simplest demodulator (detector) is a peak rectifier. The RC time constant must be short compared with the modulating period but long compared to the carrier period.

Simulation Practice

Activity *TRANSMIT*

Activity *TRANSMIT* uses the amplitude modulation circuit of Figure 13.1 to mix a high-frequency carrier with a low-frequency voice signal and transmit the result.

1. Create project *am* with schematic *TRANSMIT*.

2. Draw the modulation circuit of Figure 13.1. (*Note*: A lower-than-normal 100kHz carrier frequency is used to shorten computation times.)

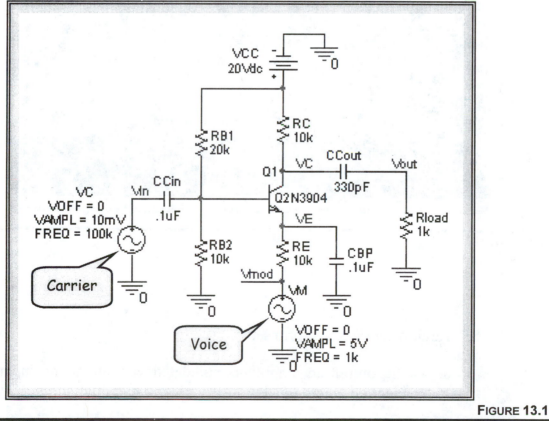

FIGURE 13.1

Amplitude modulation circuit

PSpice for Windows

3. Set the simulation profile to *Transient* from 0 to 2ms, step ceiling of 2μs, *typical* timing.

4. Run PSpice and generate the waveform set of Figure 13.2.

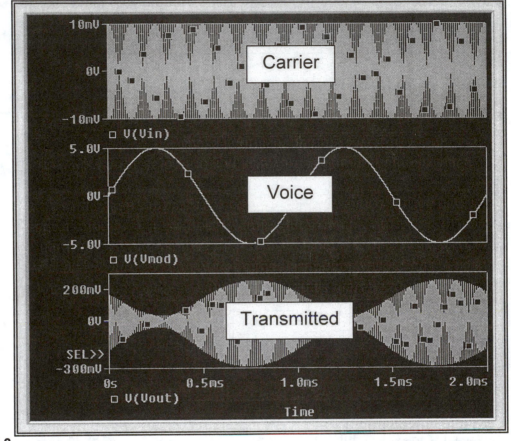

FIGURE 13.2

Amplitude modulation
waveform set

5. Reviewing the waveform set:

    a. Is the output an amplitude-modulated version of the input carrier and voice signals?

    Yes   No

b. Is the output signal maximum when the voice is minimum (negative)?

Yes   No

c. By inspection, estimate the approximate percent modulation.

modulation = _____ %

6. Reduce the audio input amplitude (V_M) to 2V, and determine by inspection the approximate reduced percent modulation. (When done, return V_M to 5V.)

Modulation ($V_M = 2V$) ≈ _____ %

Demodulation (Detection)

7. Add the amplifier/detector to your modulation circuit (Figure 13.3).

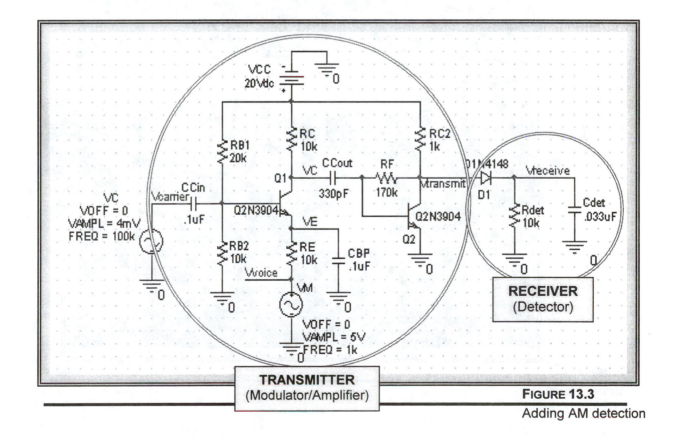

TRANSMITTER
(Modulator/Amplifier)

RECEIVER
(Detector)

FIGURE 13.3

Adding AM detection

8. If the AM transmitter/receiver works as expected, what should be the shape and frequency of the output signal ($V_{RECEIVE}$)?

    Expected shape (sine wave?) = _____

    Expected frequency = _____

9. Place an arrow at the point in the circuit where the signal would normally be transmitting into the atmosphere by way of an antenna.

10. Create the waveform set of Figure 13.4—which compares the modulation, carrier, transmit, and receive waveforms.

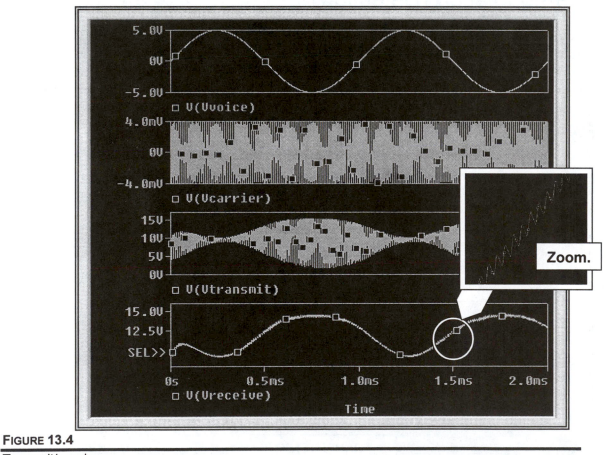

FIGURE 13.4

Transmit/receive
waveform set

11.   Referring to Figure 13.4:

a.   Does $V_{RECEIVE}$ have approximately the expected shape and frequency?

Yes   No

b.   Is the *Vreceive* output waveform actually a series of RC charge and discharge cycles whose average value follows the envelope of the carrier?

Yes   No

Fourier Analysis

12.   Perform a *FFT* (Fast Fourier Transform) analysis on the time-domain curves of Figure 13.4 to generate the frequency spectrum plots of Figure 13.5.   (Be sure to unsynchronize and expand each X-axis as shown.)

13.   Referring to Figure 13.5:

a.   Are *Vvoice* and *Vcarrier* "pure" waveforms with only a single frequency each?   (The spreading occurs because the FFT technique is only approximate.)

Yes   No

b.   Does *Vtransmit* show sidebands at plus and minus multiples of the carrier and voice frequencies?   However, are only the first harmonics (100kHz ± 1kHz) significant?

Yes   No

c.   Is the only major difference between *Vvoice* and *Vreceive* the presence of a DC component in *Vreceive*?

Yes   No

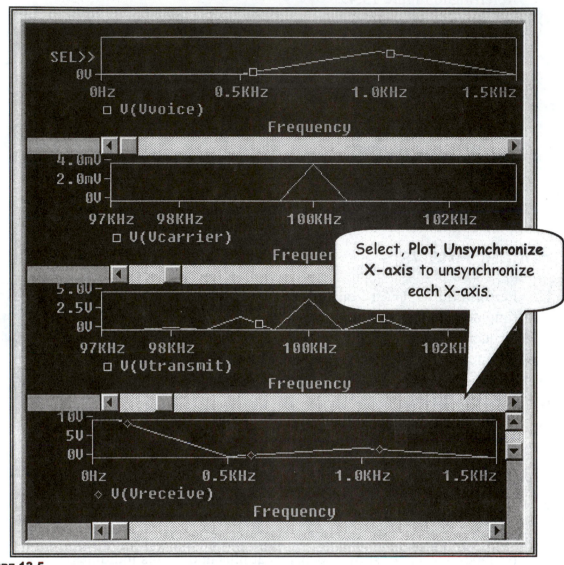

FIGURE 13.5

Frequency spectrum
waveforms

Advanced Activities

14. By examining waveforms before and after capacitor *CCout*
(Figure 13.3), what is the purpose of *CCout*? (*Hint*: What
happened to the 1kHz modulation signal?)

15. Reduce the percent modulation (Figure 13.3) and compare the resulting frequency spectrum waveform set with Figure 13.5. Explain the difference.

Exercises

1. Design an AM modulation and detection system that matches a specific station within the AM band (such as 740kHz).

2. Modulate the system with a square wave, and compare the frequency spectrum of the input and output. What does the result say about the harmonic content of a square wave and the ability of the circuit to process high modulation frequencies?

Questions and Problems

1.   Which of the following circuit elements is responsible for the gain variation during amplitude modulation?

   a. RL
   b. C1
   c. re'
   d. VCC

2.   Capacitor CBP is designed to short which of the following signals to ground?

   a. Vcarrier
   b. Vmodulation

3.   Why must the key element of a modulation circuit (such as the bipolar transistor of Figure 13.1) be a *nonlinear* device?

4.   What is the purpose of the *carrier* signal?

5.   What is the difference between a *detector* and a *peak rectifier*?

6.   For a real broadcast, why would the Fourier spectrum of the transmitted signal likely be a continuum?

7.   How does amplitude modulation provide a means of separating one radio station from another?

CHAPTER

14

Phase-Locked Loop

Phase Comparator

Objectives

- *To design and test a phase-locked loop (PLL) using the 555 timer chip*
- *To determine lock range*
- *To display the output of the phase detector*
- *To test the PLL with various incoming data streams*

Discussion

As the world gradually switches from analog to digital, one requirement is absolutely vital: All parts of the same circuit must be clocked at precisely the same frequency. (Every digital component in the same circuit must "march to the same drummer.") It sounds like no real problem—just use a common clock. However, what if the two parts of a digital circuit are a transmitter and a receiver separated by many miles?

Again, to run properly over extended periods of time, it is essential that the clock at the receiver end be precisely the same clock that controls the transmitter. The question is, How can the data and clock be mixed together, transmitted over a single line, and separated at the receiver?

The answer is to place the *phase-locked loop* (PLL) in the receiver circuit, as shown by Figure 14.1.

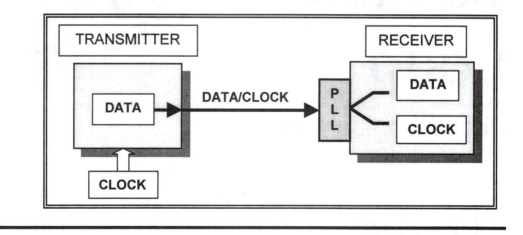

FIGURE 14.1

Initial circuit

The phase-locked loop has the amazing ability to precisely track and repeat the incoming signal—even when the transmitted data is random, arbitrary digital data. All that is necessary is for the incoming signal to have sufficient average spectral energy within the PLL's lock range.

The PLL

As shown by the block diagram of Figure 14.2, a PLL consists of three major components: a *VCO* (voltage-controlled oscillator), a *phase detector*, and a *low-pass RC filter*. It achieves its magic in the same way as many operational amplifier circuits: by negative feedback. However, with a PLL, frequency (not amplitude) is the feedback signal.

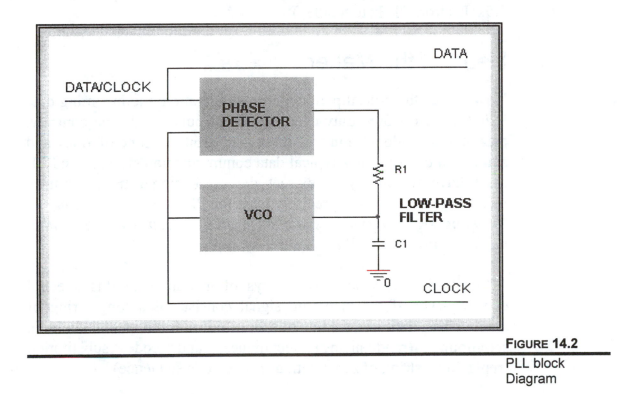

FIGURE 14.2

PLL block
Diagram

During operation the PLL works by driving the VCO output frequency to match the incoming DATA frequency.

Looking at the process in detail, the VCO output and the incoming data are both input to the phase comparator. If the incoming frequency of the data is higher or lower than the VCO frequency, the average voltage from the phase detector and low-pass filter rises or falls, which in turn drives the VCO output frequency up or down to match the incoming data frequency. The net result is that the PLL locks onto and tracks the incoming data.

PSpice for Windows

Lock range

All PLLs can maintain lock only over a limited range of frequencies. If the incoming frequency goes above or below this range, the system loses lock.

The 555-based PLL of this chapter is very crude by commercial standards and consequently has a lock range that is quite low. To be specific, we can expect to maintain lock only over approximately 1.5kHz (from 7kHz to 8.5kHz).

Heart of the Matter

Now comes the crucial part. In normal operation, the incoming data is not a perfect 50% duty cycle stream of pulses. On the contrary, data has a random nature about it—an arbitrary mix of ones and zeros. To be useful in a typical data communication system, the PLL must have the ability to extract the clock signal from arbitrary incoming data. This is precisely what the PLL does—*providing the incoming signal contains sufficient average spectral energy within the lock range of the PLL.*

In practice there are many ways of guaranteeing this spectral energy—even if the incoming signal consists of a long string of zeros. These include the use of special codes (such as Manchester), scrambling (randomizing) techniques, and code substitution (replacing a string of zeros with a special code sequence).

Low-pass Filter Memory

The final piece of the puzzle is the low-pass filter. It acts as a memory. That is, if a sudden high-frequency noise pulse briefly overrides the data signal, the capacitor voltage changes only slowly as the phase detector tries to drive the VCO output to match the noise. When the noise goes away, the capacitor "remembers" the previous VCO frequency and recovery is quick. The RC value of the filter must be high enough to provide sufficient memory yet low enough to allow proper tracking of the data frequency.

Simulation Practice

Activity *PLL*

Activity *PLL*, based on the circuit of Figure 14.3, is a phase-locked loop constructed of a 555 timer VCO. Because it is designed to lock onto square wave pulses, it is useful in data communication (digital) circuits.

1.   Create project *phaselockedloop* with schematic *PLL*.

2.   Draw the PLL circuit of Figure 14.3.   (With an ON and OFF time of .07ms, the incoming frequency is approximately 7.14kHz.)

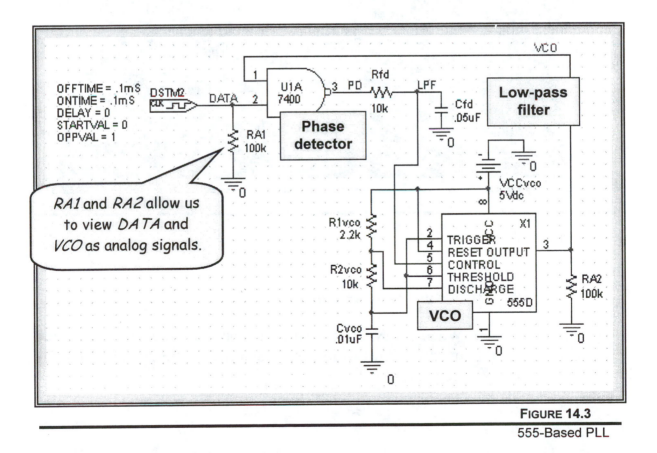

FIGURE 14.3

555-Based PLL

3. Set the simulation profile to Transient from 0 to 3ms, step ceiling of 3μs, *typical* timing.

4. Run PSpice and generate the PLL waveform set of Figure 14.4.

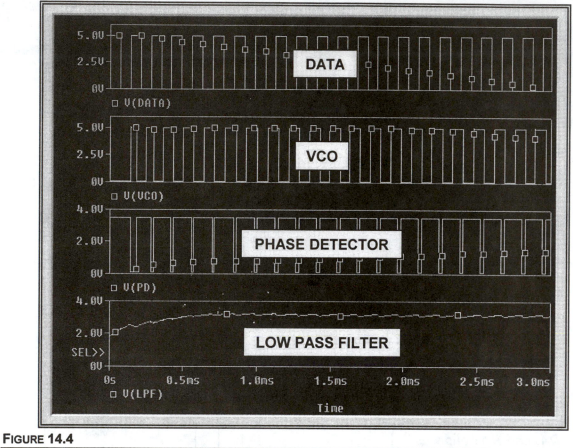

FIGURE 14.4
PLL
waveform set

5. Reviewing the waveform set:

   a. When the system is first turned on, does the VCO frequency quickly adjust to lock onto and match the incoming DATA frequency?

   Yes   No

b. When in phase lock, do the DATA and VCO signals maintain a constant phase difference?

Yes   No

c. Is the output of the phase detector high most of the time? Is the output low only when the DATA and VCO signals are both high (following NAND gate logic)?

Yes   No

d. Does the low-pass filter output show the characteristic RC rise and fall, with an approximately average voltage of 3.1V?

Yes   No

6. Change the incoming data frequency to approximately 8.33kHz (ON and OFF times of .06ms), and regenerate the waveform set of Figure 14.4.

a. Did the VCO successfully lock onto this new higher frequency?

Yes   No

b. Did the average output voltages of the phase detector and low-pass filter drop to approximately 2.8V?

Yes   No

7. Finally, set the incoming data to frequencies outside the lock range, and regenerate the waveform set of Figure 14.4. (*Hint*: 10kHz is above the lock range, and 5kHz is below the lock range.)

a. Did the PLL fail to lock onto the incoming DATA signals?

Yes   No

b. Does the irregular phase detector output (*PD*) show that the system is searching unsuccessfully for the lock state?

Yes   No

Random Data

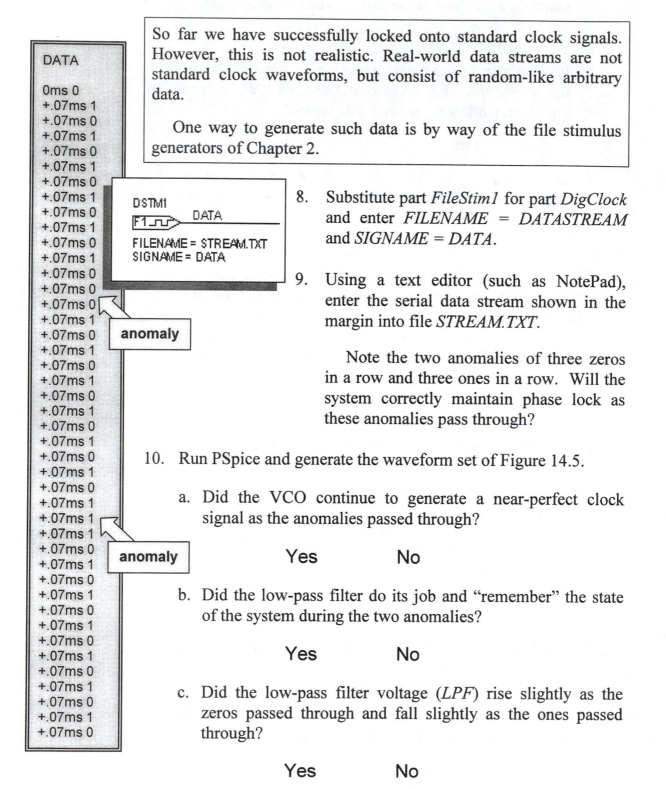

```
DATA

0ms 0
+.07ms 1
+.07ms 0
+.07ms 1
+.07ms 0
+.07ms 1
+.07ms 0
+.07ms 1
+.07ms 0
+.07ms 1
+.07ms 0
+.07ms 1
+.07ms 0
+.07ms 0
+.07ms 0
+.07ms 1
+.07ms 0
+.07ms 1
+.07ms 0
+.07ms 1
+.07ms 0
+.07ms 1
+.07ms 0
+.07ms 1
+.07ms 0
+.07ms 0
+.07ms 1
+.07ms 1
+.07ms 1
+.07ms 0
+.07ms 1
+.07ms 0
+.07ms 1
+.07ms 0
+.07ms 1
+.07ms 0
+.07ms 1
+.07ms 0
+.07ms 1
+.07ms 0
+.07ms 1
+.07ms 0
```

anomaly

anomaly

```
DSTM1
F1      DATA
FILENAME = STREAM.TXT
SIGNAME = DATA
```

So far we have successfully locked onto standard clock signals. However, this is not realistic. Real-world data streams are not standard clock waveforms, but consist of random-like arbitrary data.

One way to generate such data is by way of the file stimulus generators of Chapter 2.

8. Substitute part *FileStim1* for part *DigClock* and enter *FILENAME = DATASTREAM* and *SIGNAME = DATA*.

9. Using a text editor (such as NotePad), enter the serial data stream shown in the margin into file *STREAM.TXT*.

    Note the two anomalies of three zeros in a row and three ones in a row. Will the system correctly maintain phase lock as these anomalies pass through?

10. Run PSpice and generate the waveform set of Figure 14.5.

    a. Did the VCO continue to generate a near-perfect clock signal as the anomalies passed through?

        Yes   No

    b. Did the low-pass filter do its job and "remember" the state of the system during the two anomalies?

        Yes   No

    c. Did the low-pass filter voltage (*LPF*) rise slightly as the zeros passed through and fall slightly as the ones passed through?

        Yes   No

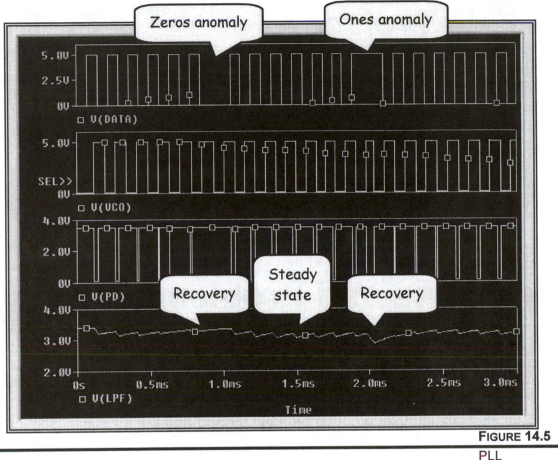

FIGURE 14.5
PLL
waveform set

Advanced Activities

11. Change the low-pass filter's RC time (up and down) and note the change in the PLL's lock ability and lock range.

12. Increase the size of the anomalies (such as four zeros or ones in a row) and note the change in operation.

13. Use other gates (such as AND, NOR, and exclusive NOR) for the phase detector and note the results.

14. Change the VCO's operating frequency, and determine the new lock range.

Exercises

1. Construct the data stream by converting part of this sentence of ASCII characters to ones and zeros. Did the system lock onto this highly arbibrary stream of data?

2. Display the DATA, VCO, and PHASE DETECTOR signals in digital rather than analog. Summarize the differences.

3. Design a PLL to lock onto signals from 9kHz to 10kHz.

Questions and Problems

1. What is a VCO?

2. Why is the PLL called a closed-loop system?

3. What signal produced by the PLL is used to clock the receiving circuit of a data communication system?

4.   Why would the PLL system lose lock upon a long series of zeros or ones?

5.   How does the PLL maintain phase lock when short noise bursts appear in the incoming data stream?

6.   Is it reasonable to assume that phase-locked loops appear at the receiving end of all network nodes?

7.   Referring to Figure 14.5 (and paying special attention to $V[PD]$), why did the low-pass filter output rise for the zeros anomaly and fall for the ones anomaly?

Part 6

Modular Design
and Applications

Whether we are planning to go to the Moon in a decade or to build the house of our dreams in a single year, one thing is clear: we must rely on top-down, modularized design practices.

We start at the top and work down the hierarchy, subdividing each task at each level, until we have a pyramid of modules—with a most general and conceptual module at the top and a base of the most detailed and specific modules at the bottom.

In the three chapters of Part 6, we demonstrate two very powerful modularization techniques available under PSpice. We finish the section and the book with two applications of modular design from the world of the telephone.

15

Modular Designs

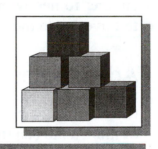

Hierarchy

Objectives

- *To break up a single-page schematic into a multipage schematic*
- *To place off-page connectors*
- *To create hierarchical designs*

Discussion

The one word that separates today's circuits from those of the past is *complexity*. Clearly, new techniques must be used when we move from a small-scale circuit of 20 components to a large-scale circuit of thousands of components.

The solution to working in any complex environment is *top-down design*. The key element in top-down design is *modularization*, in which a large-scale, complex task is broken down into a hierarchy of modules, from the general and conceptual at the top to the specific and detailed at the bottom.

So universal is top-down design that it is applied in one form or another to nearly every complex design task, from building a house to traveling to the Moon.

Advantages of Modularization

Modularizing a program into a hierarchy of modules yields a number of immediate benefits:

- Each module can be designed and displayed separately and therefore is larger and more readable.

- We can focus our attention on one module at a time, without being hindered by the complexities of the entire circuit.

- The initial design can be from a high-level perspective, in which concepts are important and low-level details can be ignored.

- We can create customized, low-level tools (sub-circuits) that can be stored in a library and used over and over (so we do not spend our time "reinventing the wheel").

- Several people can work on a given design at once.

- We can organize the design by functional parts.

To modularize a complex circuit, PSpice offers the two alternatives listed below. We implement modularization by creating multiple schematics and pages.

- Flat designs
- Hierarchical designs

Flat Designs

Flat designs are the simpler of the two and are usually reserved for small-scale circuits. Flat designs are achieved with multiple pages in a single schematic. As shown by Figure 15.1, a flat design uses *off-page connectors* to tie one page to another in a lateral (horizontal) structure.

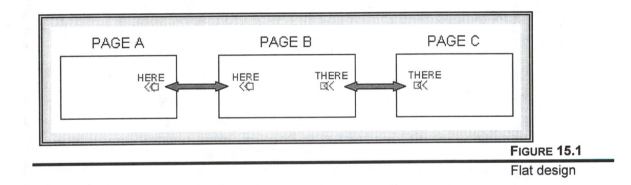

FIGURE 15.1
Flat design

Hierarchical Design

Hierarchical design handles very large and complex circuits and consists of multiple schematics, each with multiple pages. However, unlike the design previously discussed, hierarchical design is anything but flat.

To see what it means to have a hierarchy, imagine standing outside facing a large house. What you see is the overall house and its interfaces to the environment (outside doors, electrical connections, driveway, etc.). Next, walk through the front door. What you see now are a number of individual rooms. Next, walk inside the kitchen. What you now see are the various appliances and utensils for preparing meals.

What we have just done is describe a house from a top-down, hierarchical (nested) mode. In fact, that is the way houses are designed. In a hierarchical design we start at the top, with the most general and conceptual design specifications. We then subdivide the design into a pyramid of layers until we reach the most detailed and specific modules at the bottom.

As shown by Figure 15.2, a hierarchical design is just as valid for a complex circuit of seven schematics as it is for a sprawling house. Be aware that each schematic folder (A through G) can contain as many pages as it needs.

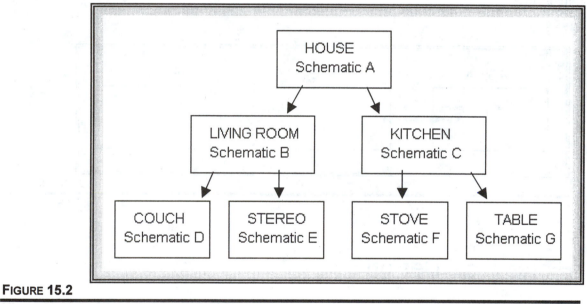

FIGURE 15.2
Hierarchical design

In this chapter, we will choose a very simple passive circuitry to illustrate the process. Our goal is to design a bandpass filter without the use of inductors. There are no tricks required; we simply cascade a high-pass and a low-pass filter and overlap their breakpoints. The peak response should then occur about halfway between their breakpoints.

Let's test our theory and learn modular design at the same time.

Simulation Practice

Activity *FLAT*

Activity *FLAT* uses two pages in a flat (horizontal) structure to design a bandpass filter. The first page (Figure 15.3) is a high-pass filter; the second (Figure 15.4) is a low-pass filter.

1. Create project *modulardesign* with schematic *FLAT*.

2. Create two pages. For the first, rename *PAGE1* to *HIGH*. For the second, **CLICKR** on the schematic name and add page *LOW*.

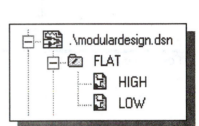

3. Into page *HIGH* enter the high-pass filter of Figure 15.3, and into page *LOW* enter the low-pass filter of Figure 15.4.

Place off-page connector

> To place the off-page connector, click the *Place off-page connector* toolbar button and choose *OFFPAGELEFT-L* for left-hand connections and *OFFPAGELEFT-R* for right-hand connections.

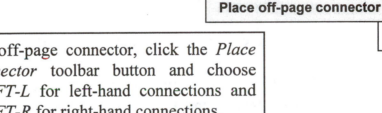

Break frequency = $1/(2\pi RC)$ = 1.59kHz

Off-page connector *OFFPAGELEFT-L*

FIGURE 15.3
High-pass filter

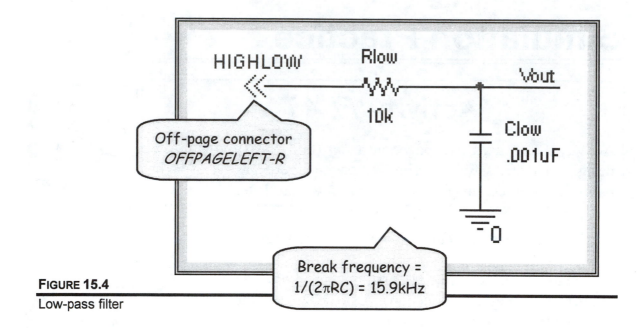

FIGURE 15.4

Low-pass filter

4. Set the simulation profile to *AC Sweep* from 1Hz to 1MEGHz, 100points/decade.

5. Run PSpice and generate the waveforms of Figure 15.5.

> The top curve shows the gain of the high-pass and low-pass filters taken separately. The bottom curve shows the overall gain of both filters.

a. Is the overall filter a bandpass filter?

Yes   No

b. Does the bottom bandpass curve seem to be a superposition of the top two individual curves?

Yes   No

c   Does the peak frequency of the bandpass filter (5.0119kHz) lie between the break frequencies of each filter?

Yes   No

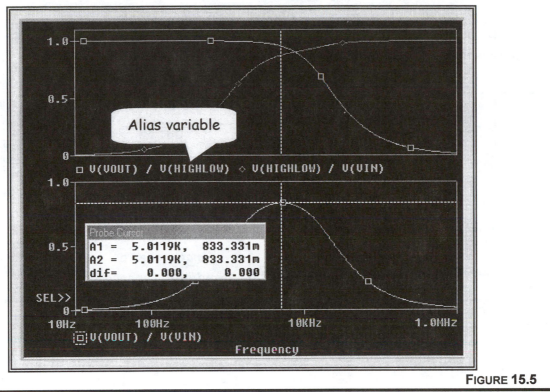

FIGURE 15.5

Bandpass
waveforms

Activity *HIERARCHY*

Activity *HIERARCHY* uses a four-block, top-down hierarchy to design our bandpass filter. Each block will correspond to a separate schematic. The four schematics will be linked together during the design.

Following the normal design sequence, we start at the top with overall block (schematic) *BANDPASS* of Figure 15.6, the most general and conceptual block.

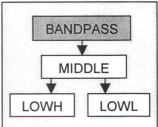

6. Create project *modulardesign*, with top-level schematic *BANDPASS*. (This schematic will remain the root schematic.)

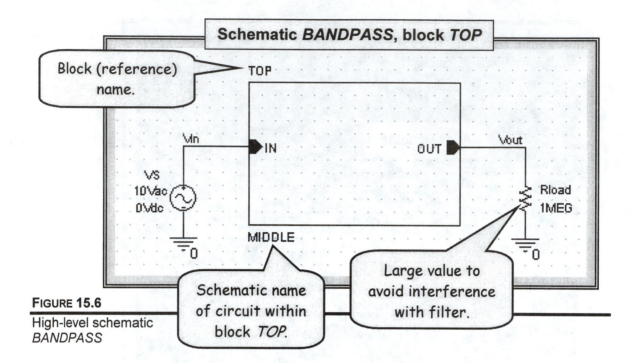

Block (reference) name.

Schematic name of circuit within block *TOP*.

Large value to avoid interference with filter.

FIGURE 15.6
High-level schematic
BANDPASS

7. To draw the high-level block diagram, perform the following steps:

a. First, we place the hierarchical block. Click the *Place hierarchical block* toolbar button to bring up the corresponding dialog box of Figure 15.7. Fill in as shown, **OK**. Drag the cursor and draw the block (as shown in Figure 15.6).

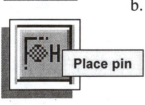

b. Next, we place the pins within the box. To accomplish this, select (**CLICKL**) the hierarchical block and click the *Place pin* toolbar button to bring up the Place Hierarchical Pin dialog box of Figure 15.8. Fill in as shown, **OK**. Drag pin to location shown, **CLICKL** to place, **CLICKR**, **End Mode**.

Again select the hierarchical block, and repeat the process for output pin *OUT* of type *output*.

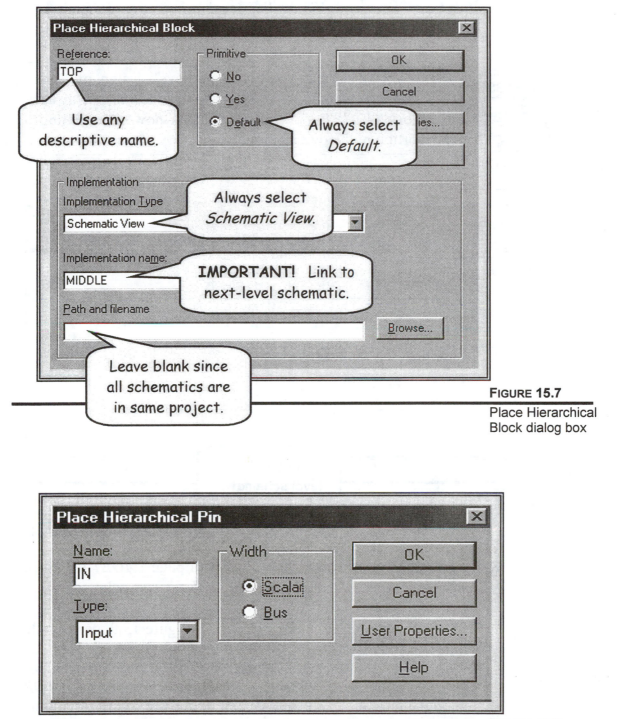

FIGURE 15.7
Place Hierarchical
Block dialog box

FIGURE 15.8
Place Hierarchical Pin
dialog box

PSpice for Windows

c. Finally, add the source and load resistor circuits and the alias tags (*Vout* and *Vin*). Top-level schematic (*TOP*) is now complete.

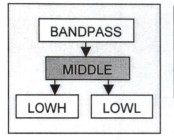

Next, we drop down to the midlevel design of Figure 15.9. As shown, it is also a block design—with each block (*HIGHPASS* and *LOWPASS*) now representing a portion of the overall circuit.

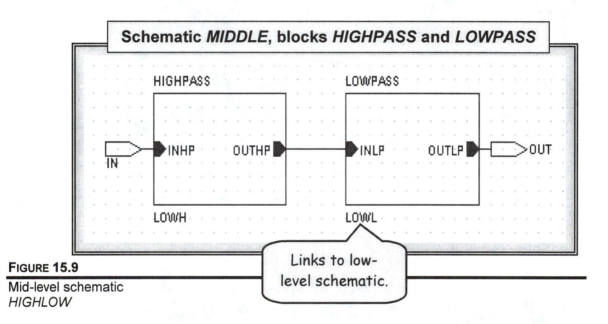

FIGURE 15.9

Mid-level schematic *HIGHLOW*

8. Add midlevel schematic *MIDDLE* with page *PAGE1* to project *modulardesign*. (Do not make *MIDDLE* the root schematic.)

9. To draw the midlevel block diagram of Figure 15.9, perform the following steps:

a. Following the same steps as with high-level schematic *TOP*, add the blocks and pins shown. As before, the reference names (*HIGHPASS* and *LOWPASS*) can be your choice; Implementation names *LOWH* and *LOWL* are the links to the next layer of schematic pages down.

b. Next, we add the two ports shown in Figure 15.9. These ports are the connection to pins *IN* and *OUT* of high-level block TOP.

Place port

To place the first (left-most) port, click the *Place port* toolbar button to bring up the Place Hierarchical Port dialog box of Figure 15.10.   Select the desired port and place as shown. (*PORTRIGHT-R* means port arrow points to the right, and the connection point is at the right.)   *Be sure to change the name (IN) to match the corresponding pin of module TOP.*

Repeat   by   placing   the   right-most   port (*PORTRIGHT-L*).

Place Hierarchical Port   ☒

Symbol:

| PORTRIGHT-R |

| PORTLEFT-R   ▲ |
| PORTNO-L |
| PORTNO-R |
| PORTRIGHT-L |
| PORTRIGHT-R   ▼ |

PORTRIGHT-R   ⊏▷∘

Libraries:

| CAPSYM |
| Design Cache |
| SOURCE |

Name:

| PORTRIGHT-R |

OK

Cancel

Add Library...

Remove Library

Help

FIGURE 15.10

Place hierarchical
Part dialog box

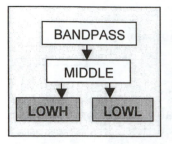

Finally, we drop down to the low-level designs of Figure 15.11. At this lowest level, the designs no longer contain blocks, but are actual circuit elements.

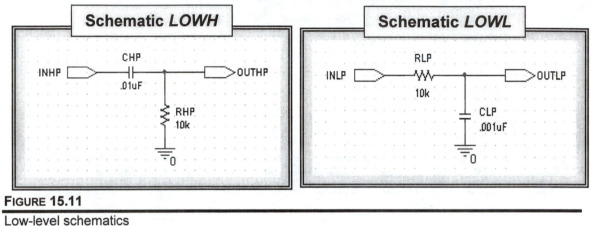

FIGURE 15.11

Low-level schematics
LOWR and *LOWL*

10. Add schematics *LOWH* and *LOWL* (each with page *PAGE1*) to project *modulardesign*.

11. Draw the circuits of Figure 15.11, making sure to match the port names to the middle-level pin names of each block.

12. The complete modular design is now done and is summarized by Figure 15.12.

13. Set the simulation profile to AC Sweep from 10Hz to 1MEGHz, 100 points/decade.

14. Run PSpice and plot *Vin* and *Vout*. Is it the same as Figure 15.5?

> Yes   No

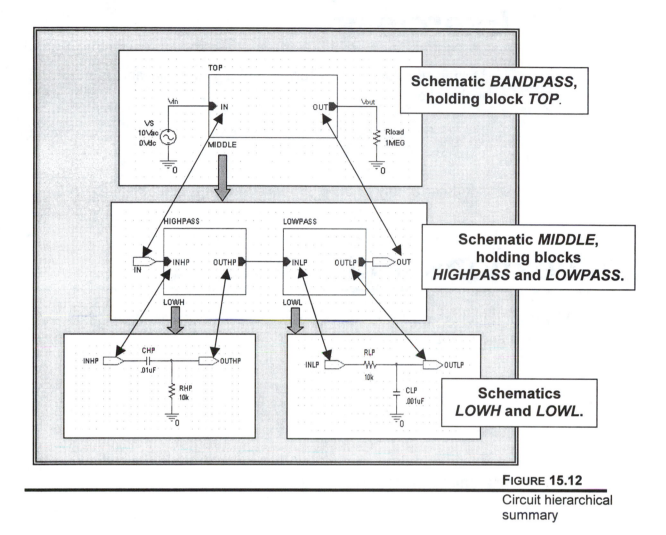

FIGURE 15.12

Circuit hierarchical
summary

Advanced Activities

15. Further modularize the bandpass filter design of Figure 15.12 by splitting up the low-pass filter into a flat arrangement of separate pages for the resistor and capacitor.

16. Repeat step 15 using hierarchy methods.

17. Modify the filter circuit of Figure 15.12 to make ground a hierarchical pin.

PSpice for Windows

Exercises

1. Design a 3-stage transistor amplifier using flat and hierarchical techniques.

2. Using hierarchical techniques modularize the PLL circuit of Chapter 14.

3. Modularize any of the previous circuits of Volumes I, II, or III.

Questions and Problems

1. Why is modular design useful for a complex project supported by several engineers and technicians?

2. Why do we normally begin designs at the top level?

3. What is the difference between a flat design and a hierarchical design?

4.   Which of the following must be a schematic name?

   Reference   Implementation name

5.   At any level a _____ in the upper-level block is linked to

   a _____ in the lower-level block.

16

Touch-Tone Decoding

Bandpass Filter

Objective

- *To use hierarchical techniques to design, test, and modify a telephone tone decoder circuit, which uses Touch-Tone Technology.*

Discussion

Our first modular design application is representative of those found in the modern-day *public switched telephone network* (PSTN). To provide the dialing function, they use *dual-tone multifrequency* (DTMF), more commonly known as *Touch-Tone Technology*.

Using a keypad similar to that of Figure 16.1, each key press generates a *pair* of tones. The two tones are summed and sent to the central office (CO), where they pass through a series of filter decoders to determine which key was pressed.

In this chapter, we will follow the action of a single key—key 6. When key 6 is pressed, tones 770Hz and 1477Hz are generated. The central office decoder circuitry must determine when key 6 was pushed and when it was not.

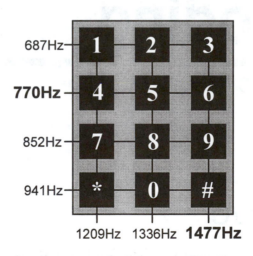

FIGURE 16.1

Tone
keypad

The tone signal will progress from our telephone to the central office and back, in a closed loop. This loop circuit is known as the *local loop*. It is typically no more than three miles in length.

Simulation Practice

Activity *LOCALLOOP*

Activity *LOCALLOOP* of this chapter will show how tones are generated by the phone and decoded by the central office. Our overall modular design will use both hierarical and flat design techniques and will follow that of Figure 16.2.

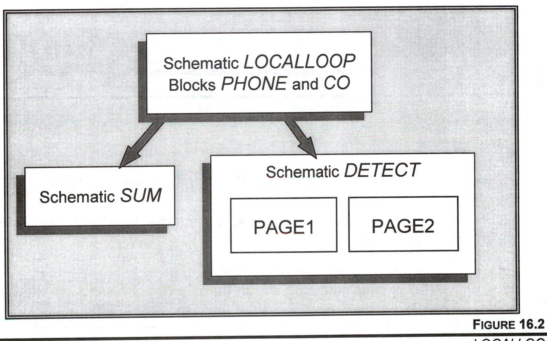

FIGURE 16.2
LOCALLOOP
modular design

1. Create project *touchtone* with schematic *LOCALLOOP*.

2. Draw the top-level schematic *LOCALLOOP* of Figure 16.3.

> The two frequencies of key 6 are summed by block *PHONE* and transmitted to block *CO* via port *LOOP* over wire *COPPER*. Within block *CO*, the combined signal passes through filters corresponding to key 6. The filter outputs are ANDed together (decoded) and presented at port *KEY6*.

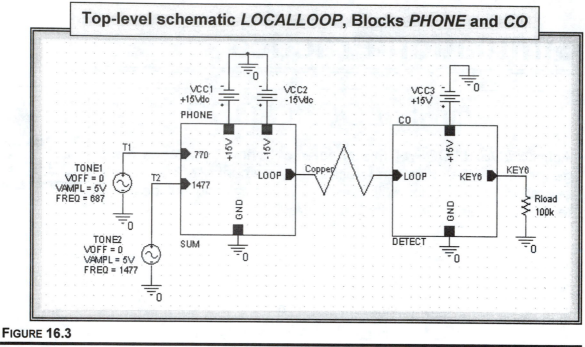

FIGURE 16.3

Touch-Tone
top-level design

3.   Add schematic *SUM* (block *PHONE*) to project *touchtone*.

4.   Using the appropriate ports, create the low-level design of Figure 16.4.

> The op amp summing circuit linearly combines the two inputs, and transmits them by way of interface port *LOOP*.

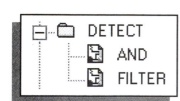

5.   Finally, we move to block CO, add schematic *DETECT* to project *touchtone*, and create the design of Figure 16.5. (Because of this block's greater complexity, we add two pages: one to handle the filter array and the second to handle the AND process.)

> Page *FILTER* accepts the LOOP signal and passes it to both behaviorally modeled bandpass filters tuned to 770 and 1477 Hz. The output is passed to page *AND*. Page *AND* accepts the two signals, ANDs them together, and passes the result to *KEY6*. A high indicates that key 6 was pressed.

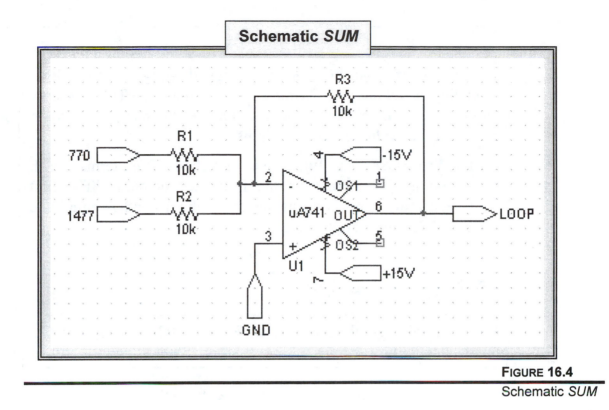

FIGURE 16.4

Schematic *SUM*

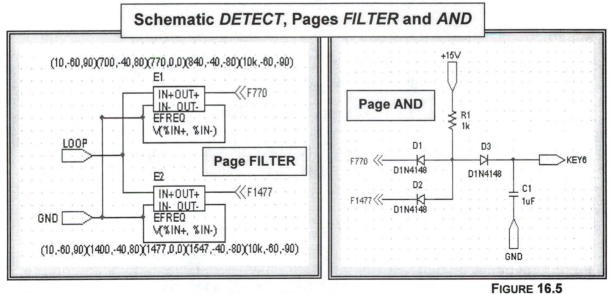

FIGURE 16.5

Schematic *DETECT*
with pages *FILTER*
and *AND*

PSpice for Windows

> ### Special Note
>
> The filters of page *FILTER* are designed using the behavioral modeling techniques of Chapter 32 (Volume II). This not only simplifies the initial design, but also reduces the component count to within the evaluation version limits.
>
> To program each filter, we specified five terms, each corresponding to a point on the frequency domain. The first number of each term is frequency (Hz), the second is amplitude (dB), and the third is phase angle.

Testing the Design

> First, we will test the *PHONE* waveforms, followed by the *CO* (central office) waveforms.

6. Set the simulation profile to *Transient* from 0 to 20ms, step ceiling of 20μs.

7. Run PSpice and display the output waveform of the *PHONE* block (Figure 16.6).

    a. Does the time-domain signal appear to be a composite of two frequencies?

    Yes   No

    b. Do the frequency components peak at approximately (within 5% of) 770Hz and 1477Hz, and are they of nearly equal amplitude?

    Yes   No

8. Next, we display the *CO* waveforms of Figure 16.7.

    Are the two filter outputs high and of nearly equal amplitude, and is the final *CO* output a hit (greater than 1.5V at steady state)?

    Yes   No

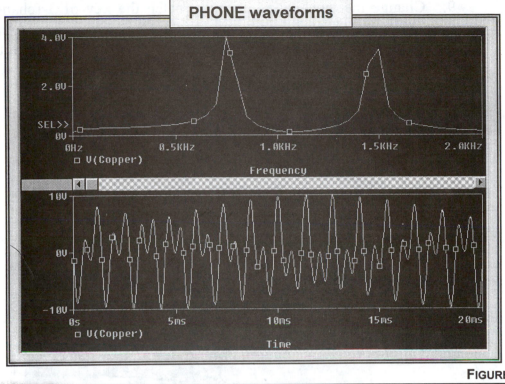

FIGURE 16.6

PHONE block
waveforms

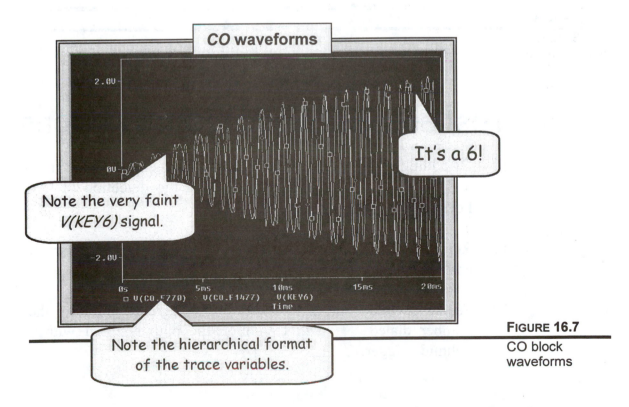

FIGURE 16.7

CO block
waveforms

9. Change the input frequencies to match the key of 3 (change 770Hz to 687Hz), and again display the output waveforms (Figure 16.8). Is the F770 output less than the F1477 output, and is the final CO output a miss (less than 1.5V)?

Yes   No

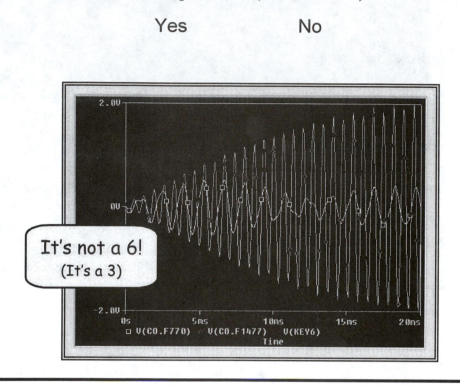

FIGURE 16.8

Key 3 output
waveforms (miss)

Test and Modification Suggestions

10. By adding an additional summing circuit branch, bandpass filter and gate, expand the system to include the detection of two keys (6 and 3?).

11. Remove the capacitor initialization from the AND gate circuit and comment on the results.

12. Feed the circuit output to a serial shift register and store the number dialed. (*Hint*: Condition the output signal with a Schmidt trigger.)

Questions and Problems

1.   Which tones are generated when a 0 is pressed?

2.   Were it not for the limitations of the evaluation version, could we have used a conventional digital AND gate in the circuit of Figure 16.5, rather than the diode-based AND gate?

3.   What function is performed by output components D1 and C1 of Figure 16.5?

4.   Why is the bandwidth of the two filters an important factor in the design?

5.   What mode of operation might prove very useful to test the characteristics of the two bandpass filters?

6. Can we run an AC Sweep analysis on the behaviorally modeled bandpass filter array of Figure 16.5?

CHAPTER

17

Pulse-Code Modulation

Time-Division Multiplexing

Objective

- *To design, test, and modify a pulse-code modulation (PCM) system.*

Discussion

In the last chapter we investigated the DTMF dialing operation between home telephone and central office. In this chapter we examine the *trunk* lines that interconnect the central offices and carry the thousands of phone signals between major cities.

For efficiency, each individual wire making up these trunk lines must carry many telephone calls simultaneously. This is accomplished by a process called *multiplexing*, the subject of this chapter.

> *Multiplexing* is the process of combining two or more signals onto a single transmission line.

FDM Versus TDM

There are two major types of multiplexing:

- FDM (frequency-division multiplexing), in which two or more signals are placed in different *frequency* slots

- TDM (time-division multiplexing), in which two or more signals are placed into different *time* slots

The older FDM system is now rapidly being replaced by the newer TDM system. When TDM includes conversion of each time-sampled analog signal to a serial digital stream, we have a PCM (pulse-code modulation) system, the subject of this chapter.

In our simplified PCM system, two input analog voice channels are continuously sampled in sequence. Each analog signal is converted to a 4-bit digital word and transmitted in serial by time multiplexing.

At the destination the process is reversed and the serial digital words are demultiplexed, converted back to analog, and output on two analog lines.

A complete PCM system includes both a transmission block and a reception block. Due to the limitations of the evaluation version, in this chapter we limit discussion to the transmission block.

Simulation Practice

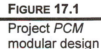

Activity *TRANSMIT*

Activity *TRANSMIT* will continuously sample two analog telephone inputs, convert them to digital, and transmit them as multiplexed, serial bit streams.

Our overall modular design is strictly hierarchical and follows the design of Figure 17.1.

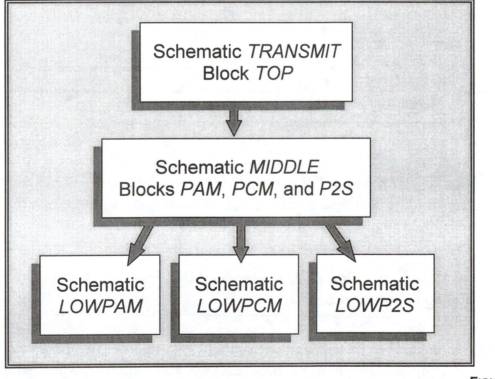

FIGURE 17.1

Project *PCM*
modular design

1.   Create project *pcm* with schematic *TRANSMIT*.

2.   Following the top-down, hierarchical process, draw the high-level circuit of Figure 17.2, a complete PCM transmitter in a single block.

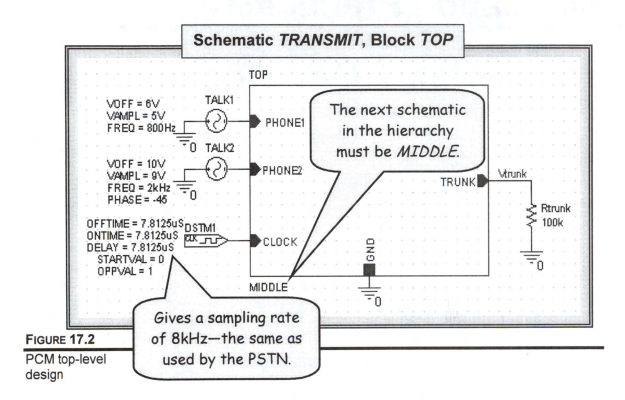

FIGURE 17.2

PCM top-level design

Inputs *TALK1* and *TALK2* represent two phones simultaneously transmitting analog signals. The two analog inputs are converted to a PCM digital serial stream by block *TRANSMIT* and sent out on line *TRUNK*.

Digital stimulus *DSTM1* provides the basic clock signal for timing all operations.

---

3.   Next, in accordance with our top-level design parameters (implementation name *MIDDLE*), we add schematic *MIDDLE* to project *pcm*.

4.   Draw the midlevel block design of Figure 17.3.

Block *PAM* (pulse-amplitude modulation) samples the analog input signals, block *PCM* (pulse-code modulation) converts each analog signal to digital, and block *P2S* (parallel to serial) converts each 4-bit digital word to serial and transmits it to output line *TRUNK*.

4 bits/word
(The PSTN uses
8 bits/word.)

Schematic *MIDDLE*, Blocks *PAM*, *PCM*, and *P2S*

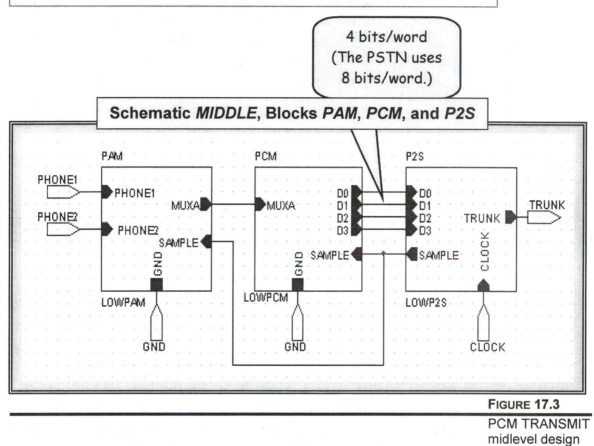

FIGURE 17.3

PCM TRANSMIT
midlevel design

5. With the middle level done, add low-level schematic *LOWPAM* to project *pcm*.

6. Draw the low-level block diagram for block *PAM* as shown in Figure 17.4.

The two analog inputs (*PHONE1* and *PHONE2*) are sampled (multiplexed) and presented at the analog output (*MUXA*). Input *SAMPLE* times the multiplex rate by successively clocking the flip-flop and shorting the input signals through M1 and M2.

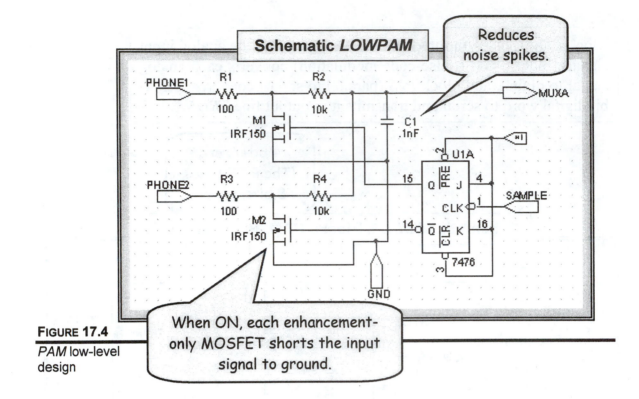

FIGURE 17.4

PAM low-level design

7. Add low-level schematic *PCMLOW* to project *pcm*, and draw the circuit of Figure 17.5.

> Each time *SAMPLE* goes active high, the multiplexed analog voltage (*MUXA*) is converted to 8-bit digital by integrated circuit (IC) *ADC8break* and presented at output lines DB0 to DB7.   Setting *VREF* to 256V calibrates the system for direct binary conversion (10V in equals 1010 out, etc.).

8. Finally, we add the remaining schematic (*LOWP2S*) to project *pcm*, and draw the circuit of Figure 17.6.

> The input parallel bits are presented to the four input lines of the 74153 dual 4-input multiplexer. In step with *CLOCK*, the *QA* and *QB* output lines of the 74163 synchronous binary counter cycle from 0 to 3 and are sent to the multiplexer's select lines. The multiplexer steers the parallel data to the single output line as the select lines are activated in order. *QB*, which goes high every fourth clock pulse, is returned to the rest of the circuit to time the sampling process.

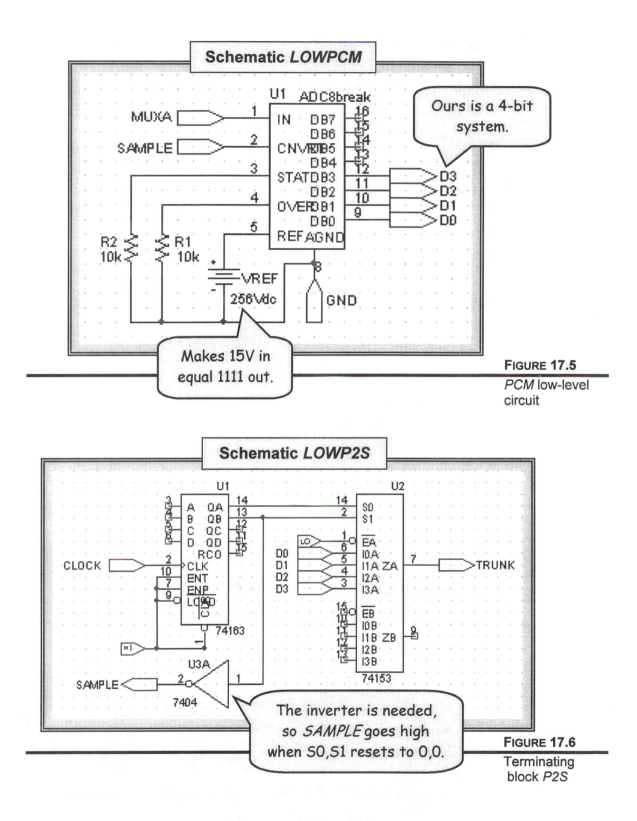

FIGURE 17.5

PCM low-level circuit

FIGURE 17.6

Terminating block *P2S*

Testing Our Design

9. Set the simulation profile to *Transient* from 0 to 1ms, step ceiling of 1µs. Be sure to initialize all flip-flops to 0 by **Options**, **Gate level simulation, 0**.

10. Run PSpice and bring up the Probe graph. Ignore for now (**Cancel**) any digital simulation errors that may occur.

To test our design, we'll move through the design from left to right, one low-level block at a time. The first block is *LOWPAM*.

---

11. Display the waveforms of block *PAM* (Figure 17.7).

    a. Is each phone sampled at 8kHz (8 times in 1 ms)?

        Yes   No

    b. Does the system toggle from *PHONE1* to *PHONE2* whenever *CLOCK* goes low?

        Yes   No

    c. Is output waveform *MUXA* a correct composite of the two input waveforms?

        Yes   No

12. Next, display the waveforms of block *PCM* (Figure 17.8).

    a. Does the digital signal (D0–D3) change when *SAMPLE* goes high?

        Yes   No

    b. Does the digital output signal approximately equal the analog in? (See the example.)

        Yes   No

    c. Would increasing the clock speed increase the accuracy?

        Yes   No

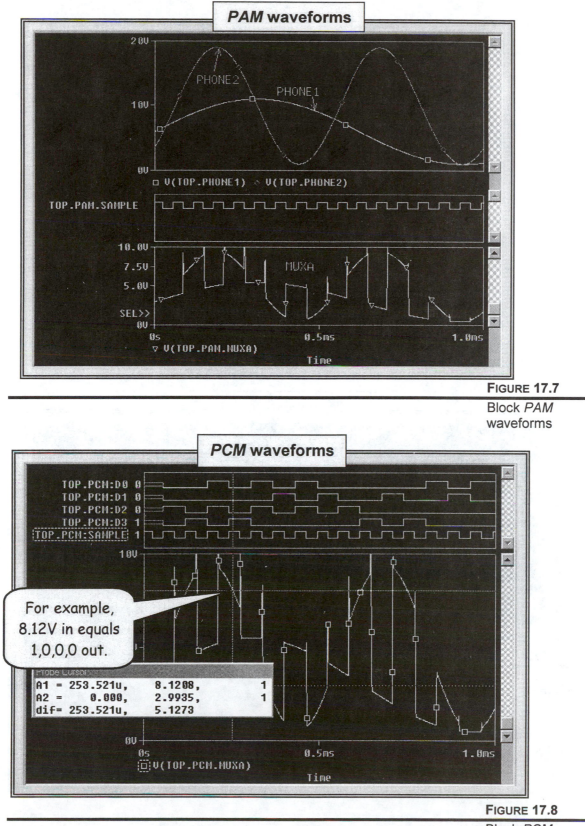

FIGURE 17.7

Block *PAM*
waveforms

FIGURE 17.8

Block *PCM*
waveforms

13. Lastly, we display the waveforms of block *P2S* (Figure 17.9). (If you wish, expand the waveforms as shown in the inset by clicking the *Zoom area* toolbar button and click the boundaries of the expansion area.)

a. Does the system sample (change) each fourth clock pulse?

Yes   No

b. Is the serial output from line *TRUNK* equal to the parallel input? (*Hint*: D0 is walked out first.)

Yes   No

c. Are the bits walked out the serial *TRUNK* line at the rate of 64k bits/sec (2 phones × 4 bits/phone × 8k samples/sec)?

Yes   No

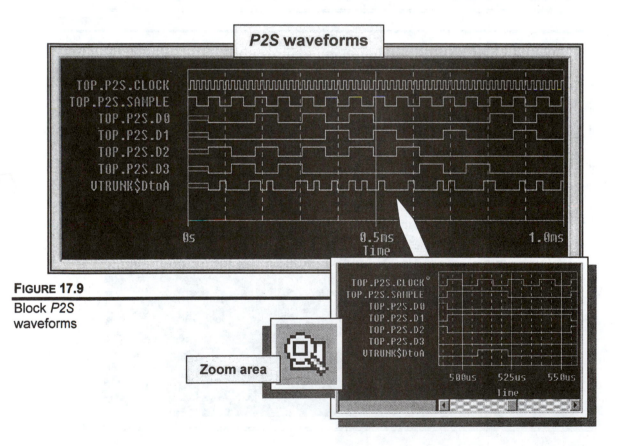

FIGURE 17.9

Block *P2S* waveforms

14. Is the overall PCM transmitter system functioning as expected?

Yes   No

Test and Modifications Suggestions

15. Perform a *worst-case* analysis on the circuit.   How would you solve the problem? (*Hint*: Most race problems are solved by adding delays at various points in the circuit.)

16. Design and test the *RECEIVE* block of our PCM system. (Hint: Reverse the *PAM*, *PCM*, and *P2S* processes.) To avoid exceeding the limitations of the evaluation version, perform this as a separate project.

17. Expand the system to multiplex four phones.

18. Redesign the system for 8-bit operation.

19. Use a 74194 shift register, instead of the multiplex, to carry out the *P2S* block.

20. Redesign the *PAM* block of Figure 17.4 to eliminate crosstalk. (*Hint*: Make use of an ICVS op amp.)

Questions and Problems

1. Why are we guaranteed that switches M1 and M2 of block *PAM* can never be ON at the same time?

2. Does the system obey the *Nyquist* theorem? (Is the sampling rate for both PHONE1 and PHONE2 at least twice the input frequency?)

3. Following up question two, what is the highest analog input frequency that the circuit of Figure 17.2 can sample without losing data?

4. Why does TDM (time-division multiplexing) tend to produce noise spikes?

5. The basic PCM multiplexing unit for the phone system consists of 24 phones. If each phone is sampled at the 8kHz rate and each analog voltage is converted to 8-bit digital, what is the serial output bit rate from this basic unit? (Why do you think it is slightly less than the telephone system's 1.544MHz T1 rate?)

6. With our system the transmitter and receiver were synchronized with a separate clock line. In the real phone system, there is no separate clock line. How, then, does the phone system remain synchronized? (*Hint*: See Chapter 14.)

7. If the transmission block consists of a multiplexer, analog-to-digital converter, and parallel-to-serial converter, what should the receive block consist of?

Appendix A

Simulation Notes Summary

Volume I

The starred notes (✱) are repeated here in Appendix A.
For the others, you must refer directly to Volume I.

Note	Title
1.1 ✱	How do I create a new project?
1.2 ✱	How do I place parts on the schematic?
1.3 ✱	How do I "refresh" or "resize" my circuit?
1.4 ✱	How do I reposition components?
1.5 ✱	How do I change value attributes?
1.6 ✱	How do I set the simulation profile?
1.7	How do I determine the direction of conventional current?
2.1	How do I label wires (nets)?
2.2	How do I add another schematic to a project?
2.3	How do I make a schematic the root schematic?
2.4	How do I reset the root schematic and the active profile?
3.1	How do I change a part name?
5.1	How do I use markers?
6.1	How do I make a component value a variable?
6.2	How do I display or modify a trace?
7.1	Why are sinusoidal waveforms so special?
7.2	How do I create a second Y-axis?
7.3	How do I change the axis settings?
7.4 ✱	How do I use the cursor?
7.5	How does PSpice perform transient calculations?
7.6	How do I print my schematics and plots?
8.1	How do I create multiple plots?
12.1	How do I change the axis settings?
13.1	What special markers are available to plot advanced waveforms?
13.2	How do I open up multiple output windows?
15.1	How do I expand or compress waveforms?
16.1	How do I mark coordinate values on my graphs?
16.2	How do I display the properties of a waveform?
17.1	How do I combine waveforms from different schematics?

Volume II

Note	Title
1.1	How do I change the X-axis variable?
1.2	How do I change model parameters?
2.1	How do I change the ambient temperature?
2.2	How do I set up the DC Sweep nested mode?
5.1	How do I uncouple plots?
5.2	How do I set a watch alarm?
10.1	How do I generate a damped sine wave?

Volume III

Note	Title	Page
1.1	How do I expand digital waveforms?	15
1.2	How do I change the size of the digital display?	17
2.1	How do I add a system bus?	31
4.1	How do I set the mode of timing?	57

Simulation Note 1.1
How do I create a new project?

1) **CLICKL** the Create Document toolbar button (or **File**, **New**, **Project**) to bring up the New Project dialog box, fill in as shown, **OK**.

2) Select *Create a blank project* in the Create PSpice Project dialog box, **OK** to open the Capture window.

Simulation Note 1.2
How do I place parts on the schematic?

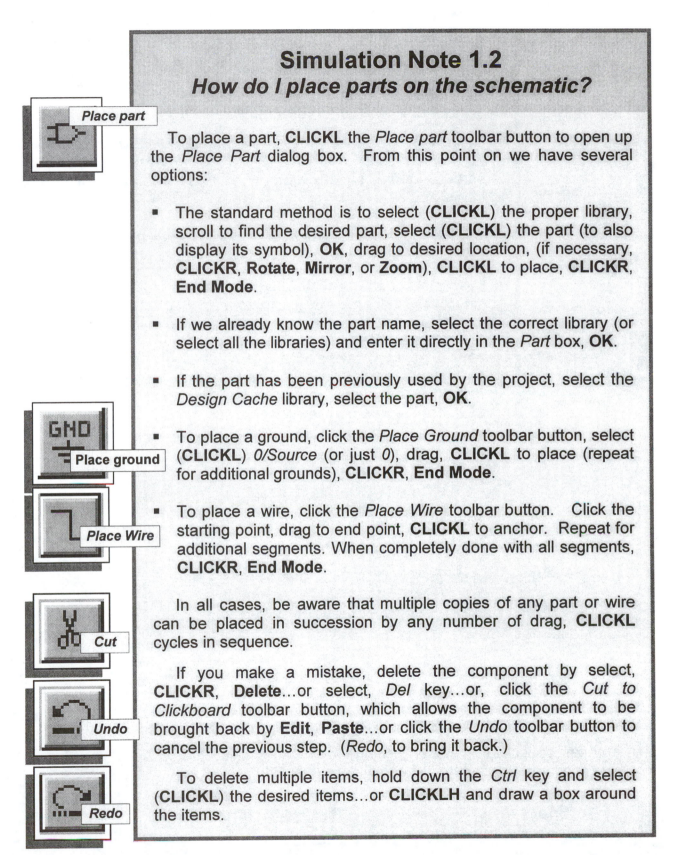

Place part

To place a part, **CLICKL** the *Place part* toolbar button to open up the *Place Part* dialog box. From this point on we have several options:

- The standard method is to select (**CLICKL**) the proper library, scroll to find the desired part, select (**CLICKL**) the part (to also display its symbol), **OK**, drag to desired location, (if necessary, **CLICKR**, **Rotate**, **Mirror**, or **Zoom**), **CLICKL** to place, **CLICKR**, **End Mode**.

- If we already know the part name, select the correct library (or select all the libraries) and enter it directly in the *Part* box, **OK**.

- If the part has been previously used by the project, select the *Design Cache* library, select the part, **OK**.

Place ground

- To place a ground, click the *Place Ground* toolbar button, select (**CLICKL**) *0/Source* (or just *0*), drag, **CLICKL** to place (repeat for additional grounds), **CLICKR**, **End Mode**.

Place Wire

- To place a wire, click the *Place Wire* toolbar button. Click the starting point, drag to end point, **CLICKL** to anchor. Repeat for additional segments. When completely done with all segments, **CLICKR**, **End Mode**.

In all cases, be aware that multiple copies of any part or wire can be placed in succession by any number of drag, **CLICKL** cycles in sequence.

Cut

If you make a mistake, delete the component by select, **CLICKR**, **Delete**...or select, *Del* key...or, click the *Cut to Clickboard* toolbar button, which allows the component to be brought back by **Edit**, **Paste**...or click the *Undo* toolbar button to cancel the previous step. (*Redo*, to bring it back.)

Undo

To delete multiple items, hold down the *Ctrl* key and select (**CLICKL**) the desired items...or **CLICKLH** and draw a box around the items.

Redo

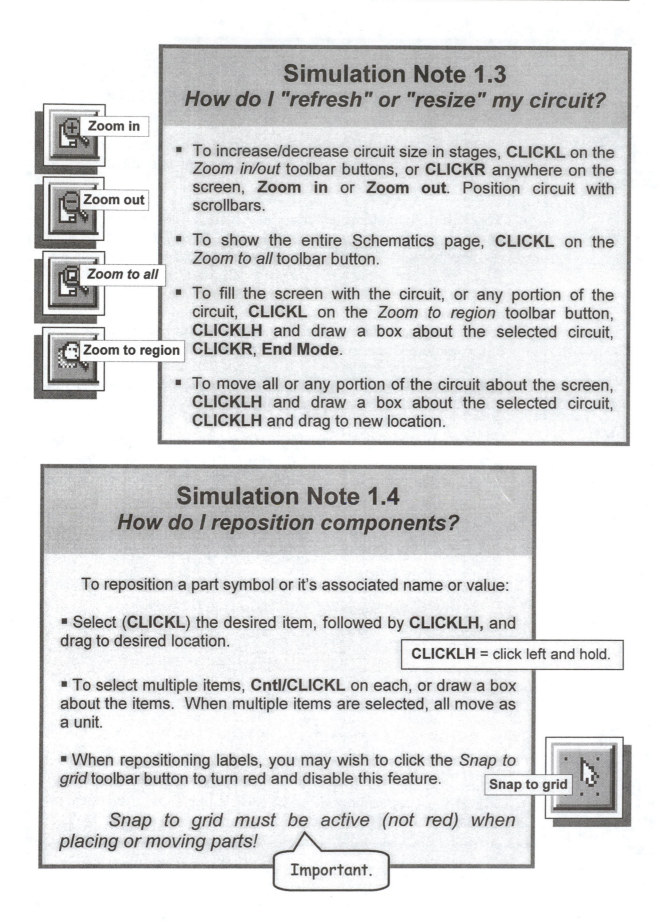

Simulation Note 1.3
How do I "refresh" or "resize" my circuit?

Zoom in

Zoom out

Zoom to all

Zoom to region

- To increase/decrease circuit size in stages, **CLICKL** on the *Zoom in/out* toolbar buttons, or **CLICKR** anywhere on the screen, **Zoom in** or **Zoom out**. Position circuit with scrollbars.

- To show the entire Schematics page, **CLICKL** on the *Zoom to all* toolbar button.

- To fill the screen with the circuit, or any portion of the circuit, **CLICKL** on the *Zoom to region* toolbar button, **CLICKLH** and draw a box about the selected circuit, **CLICKR**, **End Mode**.

- To move all or any portion of the circuit about the screen, **CLICKLH** and draw a box about the selected circuit, **CLICKLH** and drag to new location.

Simulation Note 1.4
How do I reposition components?

To reposition a part symbol or it's associated name or value:

- Select (**CLICKL**) the desired item, followed by **CLICKLH,** and drag to desired location.

> **CLICKLH** = click left and hold.

- To select multiple items, **Cntl/CLICKL** on each, or draw a box about the items. When multiple items are selected, all move as a unit.

- When repositioning labels, you may wish to click the *Snap to grid* toolbar button to turn red and disable this feature.

Snap to grid

Snap to grid must be active (not red) when placing or moving parts!

> Important.

Simulation Note 1.5
How do I change value attributes?

To change a value attribute:

   DCLICKL on the attribute to select and bring up the *Display Properties* dialog box. Enter the correct value in the *Value* field, select the desired display format, **OK**. If necessary, click the *snap to grid* button and reposition the attributes.

   (For most parts, the default *Value Only* display format is satisfactory.   Had we selected *Name and Value* for *V1* and *R1*, then *DC=10V* and *Value=5k* would have been displayed.)

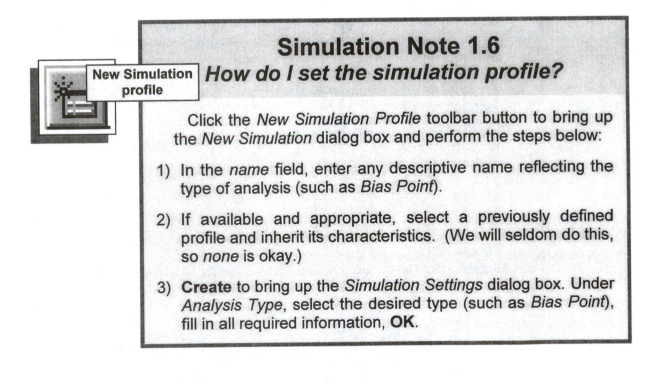

Simulation Note 1.6
How do I set the simulation profile?

New Simulation profile

   Click the *New Simulation Profile* toolbar button to bring up the *New Simulation* dialog box and perform the steps below:

1) In the *name* field, enter any descriptive name reflecting the type of analysis (such as *Bias Point*).

2) If available and appropriate, select a previously defined profile and inherit its characteristics. (We will seldom do this, so *none* is okay.)

3) **Create** to bring up the *Simulation Settings* dialog box. Under *Analysis Type*, select the desired type (such as *Bias Point*), fill in all required information, **OK**.

Simulation Note 7.4
How do I use the cursor?

To activate the cursors, **CLICKL** on the *Toggle Cursor* toolbar button. Note the appearance of the cursors and the Cursor window. If necessary, reposition the Cursor window (**CLICKLH** on the color bar and drag).

There are two cursors, A1 and A2.

- A1 is associated with closely spaced dotted lines and is controlled by the left-hand mouse button.

- A2 is associated with loosely spaced dotted lines and is controlled by the right-hand mouse button.

Cursor use is as simple as 1, 2, and 3.

1. To associate a cursor with a waveform, **CLICKL** (for cursor A1) or **CLICKR** (for cursor A2) on the appropriate color-coded legend symbols in front of the trace variables along the bottom of the graph.

2. To position either cursor, **CLICKL** or **CLICKR** at any graph location and cursor A1 or A2 moves to that horizontal location. (**CLICKLH** or **CLICKRH** to drag either cursor). The *x/y* coordinates of each cursor (as well as the difference) appear in the cursor window.

3. To fine tune A1 (or A2), use the arrow keys (or Shift arrow keys).

To quickly move the cursor to certain key points, click any of the corresponding toolbar buttons shown below (or **Trace**, **Cursor**, select option).

View, Toolbars, Cursor to bring up.

| Peak | Trough | Slope | Min | Max | Point |

To deactivate the cursor, **CLICKL** on *Toggle cursor* toolbar button.

Appendix B
Axis and Grid Control

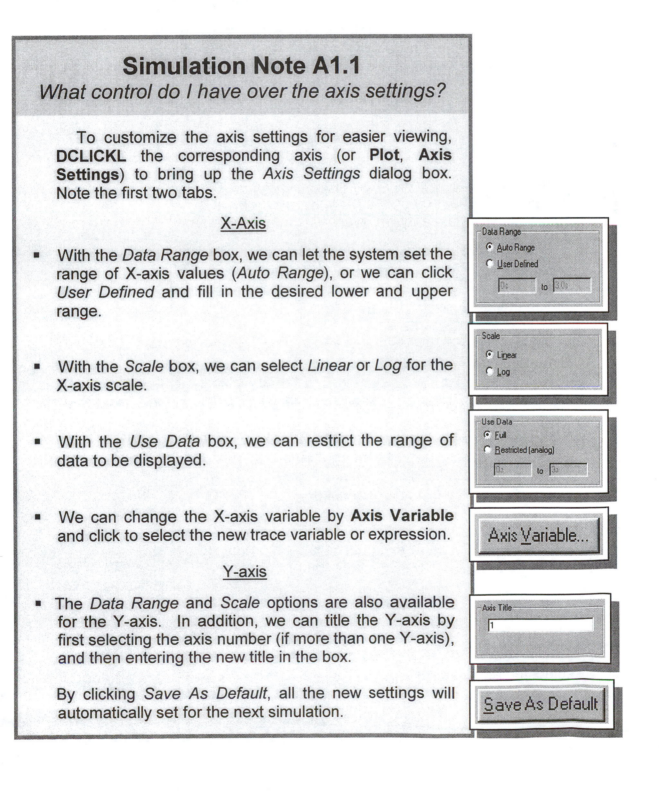

Simulation Note A1.1
What control do I have over the axis settings?

To customize the axis settings for easier viewing, **DCLICKL** the corresponding axis (or **Plot, Axis Settings**) to bring up the *Axis Settings* dialog box. Note the first two tabs.

X-Axis

- With the *Data Range* box, we can let the system set the range of X-axis values (*Auto Range*), or we can click *User Defined* and fill in the desired lower and upper range.

- With the *Scale* box, we can select *Linear* or *Log* for the X-axis scale.

- With the *Use Data* box, we can restrict the range of data to be displayed.

- We can change the X-axis variable by **Axis Variable** and click to select the new trace variable or expression.

Y-axis

- The *Data Range* and *Scale* options are also available for the Y-axis. In addition, we can title the Y-axis by first selecting the axis number (if more than one Y-axis), and then entering the new title in the box.

By clicking *Save As Default*, all the new settings will automatically set for the next simulation.

Simulation Note A1.2
How do I change the grid settings?

To customize the grid settings for easier viewing, DCLICKL the appropriate axis (or **Plot**, **Axis Settings**) to bring up the *Axis Settings* dialog box. Note the last two tabs.

X Grid

The *Major* section controls the solid grid lines and the placement of X-axis value; the *Minor* box controls the dotted lines between the major lines. When *Automatic* is selected, the major and minor grid spacing is automatically chosen by the system.

- With *automatic* disabled, the *Spacing* box allows us to set the major (solid) grid lines and their associated values. For example, a linear value of 1 places the lines and values at 1, 2, 3, etc.).

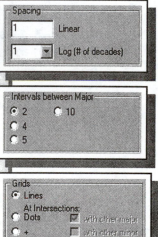

The *Invervals between Major* box sets the number of dotted line segments that fit between each major range. For example, selecting 2 intervals places a single minor dotted line between each major solid line.)

- With the *Grids* box we can enable (**Lines**) or disable (**None**) the display of grid lines (both major and minor). We can also specific that dots or crosses appear at the Intersection of major and minor lines.

- Finally, we can place the *ticks* (the small markers that appear along the axis) inside or outside the plot. (only the major grids have associated numbers.)

Y-axis

- The Y-axis grid has the same controls as the X-axis. Just remember if there is more than one Y-axis to first select the axis number.

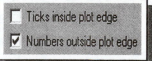

Appendix C

Specification Sheets

LM741
OPERATIONAL AMPLIFIER

Parameter	Conditions	Value			Unit
		Min	Typ	Max	
Input Offset Voltage	TA = 25°C		1.0	5.0	mV
Input Offset Current	TA = 25°C		20	200	nA
Input Bias Current	TA = 25°C		80	500	nA
Input Resistance	TA = 15°C, VS = ± 20V	.3	2.0		Mohms
Large Signal Voltage Gain	TA = 25°C, VS = ± 15V	50	200		V/mV
Output Short Circuit Current	TA = 25°C		25		mA
Common-Mode Rejection Ratio		70	90		DB
Bandwidth	TA = 25°C	.437	1.5		MHz
Slew Rate	TA = 25°C, Unity Gain		.5		V/us
Supply Current	TA = 25°C		1.7	2.8	mA
Power Consumption	TA = 25°C, VS = ± 15V		60	100	mW

TTL FAMILY CHARACTERISTICS

Standard 54/74

Parameter	Test Conditions	Min	Typ	Max	Unit
VIH Input HIGH voltage	Guaranteed input HIGH voltage for all inputs	2.0			V
VIL Input LOW voltage	Guaranteed input LOW voltage for all inputs			.8	V
VCD Input Clamp Diode Voltage	VCC = Min Iin = −12mA		−0.8	−1.5	V
VOL Output LOW voltage	VCC = Min IOL = 16mA			0.4	V
VOH Output HIGH voltage	VCC = Min IOH = −800uA	2.4	3.5		V
IOH Output HIGH current (open collector)	VCC = Max Vout = 5.5V			250	uA
IOZH Output "off" current HIGH (3 state)	VCC = Max Vout = 2.4V VOE = 2.0V			40	uA
IOZL Output "off" current LOW	VCC = Max Vout = .5V VOE = 2.0V			−40	uA
IIH Input HIGH current	VCC = Max Vin = 2.4V			40	uA
IIH Input HIGH current at max input voltage	VCC = Max Vin = 5.5V			1.0	mA
IIL Input LOW current	VCC = Max Vin = 0.4V			−1.6	mA
IOS Output short circuit current	VCC = Max Vout = .0V	−18		−55	mA

Low- Power Schottky 54LS/74LS

Parameter	Test Conditions	Min	Typ	Max	Unit
VIH Input HIGH voltage	Guaranteed input HIGH voltage for all inputs	2.0			V
VIL Input LOW voltage	Guaranteed input LOW voltage for all inputs			.8	V
VCD Input Clamp Diode Voltage	VCC = Min Iin = −12mA		−0.65	−1.5	V
VOL Output LOW voltage	VCC = Min IOL = 16mA			0.4	V
VOH Output HIGH voltage	VCC = Min IOH = −800uA	2.7	3.4		V
IOH Output HIGH current (open collector)	VCC = Max Vout = 5.5V			100	uA
IOZH Output "off" current HIGH (3 state)	VCC = Max Vout = 2.4V VOE = 2.0V			20	uA
IOZL Output "off" current LOW	VCC = Max Vout = .5V VOE = 2.0V			−20	uA
IH Input HIGH current	VCC = Max Vin = 2.4V			20	uA
II Input HIGH current at max input voltage	VCC = Max Vin = 5.5V			0.1	mA
IIL Input LOW current	VCC = Max Vin = 0.4V			−0.46	mA
IOS Output short circuit current	VCC = Max Vout = .0V	−18		−100	mA

7442A BCD-to-Decimal Decoder

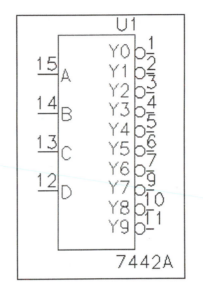

A3	A2	A1	A0	0	1	2	3	4	5	6	7	8	9
L	L	L	L	L	H	H	H	H	H	H	H	H	H
L	L	L	H	H	L	H	H	H	H	H	H	H	H
L	L	H	L	H	H	L	H	H	H	H	H	H	H
L	L	H	H	H	H	H	L	H	H	H	H	H	H
L	H	L	L	H	H	H	H	L	H	H	H	H	H
L	H	L	H	H	H	H	H	H	L	H	H	H	H
L	H	H	L	H	H	H	H	H	H	L	H	H	H
L	H	H	H	H	H	H	H	H	H	H	L	H	H
H	L	L	L	H	H	H	H	H	H	H	H	L	H
H	L	L	H	H	H	H	H	H	H	H	H	H	L

All others   All high

7474 Dual D-Type FLIP-FLOP

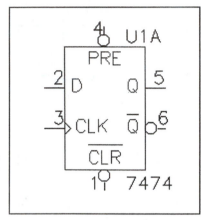

OPERATING MODE	INPUTS				OUTPUTS	
	SD	RD	CP	D	Q	NQ
Asynchronous Set	L	H	X	X	H	L
Asynchronous Reset	H	L	X	X	L	H
Undetermined	L	L	X	X	H	H
Load "1" (Set)	H	H	Rise	H	H	L
Load "0" (Reset)	H	H	Rise	L	L	H

7476 Dual JK Flip-Flop

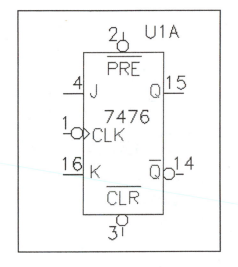

OPERATING MODE	INPUTS					OUTPUTS	
	SD	RD	CP	J	K	Q	NQ
Asynchronous Set	L	H	X	X	X	H	L
Asynchronous Reset	H	L	X	X	X	L	H
Undetermined	L	L	X	X	X	H	H
Toggle	H	H	Pulse	H	H	NQ	Q
Load "1" (Set)	H	H	Pulse	H	L	H	L
Load "0" (Reset)	H	H	Pulse	L	H	L	H
Hold (No Change)	H	H	Pulse	L	L	Q	NQ

7485 4-Bit Magnitude Comparator

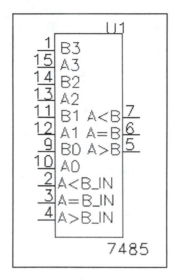

Input signals	A>B	A=B	A<B
A>B	H	L	L
A=B	L	H	L
A<B	L	L	H

7493A 4-Bit Binary Ripple Counter

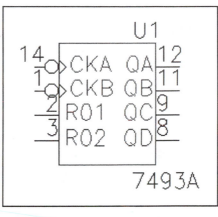

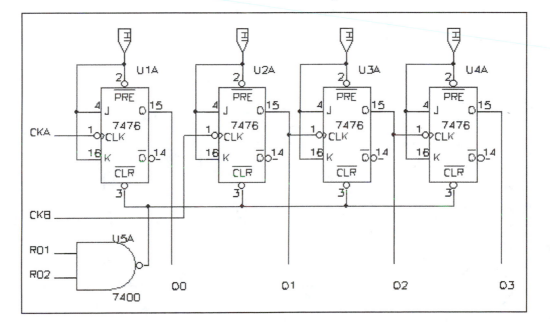

RESET INPUTS		OUTPUTS			
R01	R02	Q0	Q1	Q2	Q3
H	H	L	L	L	L
L	H		Count		
H	L		Count		
L	L		Count		

74147 10-Line-to-4-Line Priority Encoder

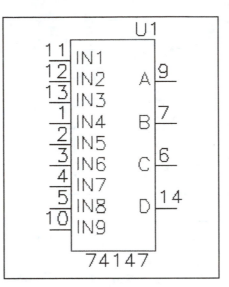

i1	I2	I3	I4	I5	I6	I7	I8	I9	A3	A2	A1	A0
H	H	H	H	H	H	H	H	H	H	H	H	H
X	X	X	X	X	X	X	X	L	L	H	H	L
X	X	X	X	X	X	X	L	H	L	H	H	H
X	X	X	X	X	X	L	H	H	H	L	L	L
X	X	X	X	X	L	H	H	H	H	L	L	H
X	X	X	X									
X	X	X										
X	X											
X												
L	H	H	H	H	H	H	H	H				

Index